DROWNING IN POTENTIAL

HOW AMERICAN SOCIETY CAN SURVIVE DIGITAL TECHNOLOGY

ROD WALLACE

Published by Economic Systems Press

Cover design by Ida Fia Sveningsson at IdaFiaSveningsson.se
Interior design by Matías Baldanza
Cover Figures & Illustrations by author K. Limkin, licensed under Creative
Commons license.
Illustrations in the text by Marlon Brando Gonzales
at getyourbookillustrations.com.

ISBN-10: 1-7326240-0-3
ISBN-13: 978-1-7326240-0-9

Printed in the United States of America

To my family:
My source of
inspiration, love,
and desire for
a bright future.

ABOUT THE AUTHOR

Rod Wallace, PhD, is a leading business strategist and speaker focused on energizing organizations to deliver profit and make a difference.

With a unique background combining international business, economics, and technology, Wallace challenges teams to integrate diverse perspectives into a coherent purpose with value for both investors and society. Wallace recent research provides groundbreaking insights, strategies, and tools for overcoming our society's most pressing issues.

Organizations choose to work with Wallace because of his innovative approach. He challenges businesses to deliver the greatest profit by expanding their ability to improve society.

Wallace guides companies and executives from startups to Fortune 500 toward concrete solutions to large-scale problems. Global leaders working with Wallace have delivered strategic plans for businesses with over $10 billion in turnover and strategic fit analyses for more than $4 billion in potential acquisitions.

Wallace served as a Leadership Team Member for Cargill Incorporated Refined Oils Europe, a $3 billion enterprise head-quartered in the Netherlands. He collaborated with Silicon Valley pioneer Dr. Steve Omohundro in exploring the impact of artificial intelligence and other cutting-edge technology on society. And he was an invited researcher to the Japanese Ministry of Economy, Trade, and Industry.

ABOUT THE AUTHOR

Wallace earned his PhD from the University of Michigan. A Fulbright Fellow, he studied with leading economic historian Gary Saxonhouse. Wallace's academic research and publications have focused on the interplay between business collaboration and competition.

Wallace has lived on four continents and is currently based in Wilmington, Delaware. As a husband and the father of three young sons, he worries about the state of the world we will pass on to the next generation. Wallace believes American society must do more than survive—it must flourish.

For more about the author,
visit RodWallacePhD.com

CONTENTS

INTRODUCTION

Three hours after being admitted to New York Central Hospital, 78-year-old Betty lies on a gurney in an emergency room corridor. Only a sheet covers Betty and she is shivering. Betty was admitted with a potentially life-threatening infection but hasn't seen a doctor yet. She is terrified and doesn't know if anyone has called her son. Too flustered to flag one of the medical staff people who scurry by, Betty lies there silently, praying that someone might acknowledge her. She practices the words to request a blanket but speaks them to no one.

On the other side of the ER, Nurse Nancy glances down at her pager that is calling her to deliver medication to Patient 18-532. She clips the pager to her scrubs, then with cell phone in her pocket and digital tablet tucked under her arm, Nancy begins her trek to Room 302 where Patient 18-532 should be. But Betty's not there. Nancy's colleague Anna settled Betty in the corridor instead: Room 302 wasn't ready, and Anna had other patients to see. In her hurry, Anna didn't even think about updating Betty's chart with her new location.

Already feeling short-staffed and overworked, Nurse Nancy searches across the ER for Betty. With every second away from her desk, Nancy feels the weight of all the incomplete paperwork waiting on her computer. The hospital recently introduced a new platform for keeping patient records and Nancy does not yet feel comfortable with it.

Nancy finds Betty in the hallway and delivers the medication with a brisk smile. While she updates Betty's chart on her tablet, Nancy's cell phone rings with an emergency and she dashes off to her next task before Betty can ask for a blanket. Nancy became a nurse to care for others, but due to the frenetic schedule and digital demands that have become her norm, she did not get a chance to think about connecting with Betty on a personal level.

Nancy, Anna, and their colleagues have access to the most cutting-edge medical technology. Yet, they failed Betty at the most basic, human level. Superficially, this scene seems independent of technology. Two nurses failed to provide warm, humanizing care to a patient. This perspective is true to an extent.

Nonetheless, the impact of Digital Technology (D-Tech) lurks just below the surface. Technology has changed the nature of Nurse Nancy's challenges. One hundred and fifty years ago, Nancy would have worked with a single, generalist physician and a handful of notes on paper. Her medical knowledge would have been limited, but she would have gotten to know the patient and paid attention to their comfort.

As technology has improved, Nancy's world has become more complex. Despite Nancy's years of training, she is an expert on only a sliver of modern medical technology. Large teams of healthcare workers with different specialties are now required to take care of an individual patient. Rather than a handful of notes, Nurse Nancy has access to hundreds of electronic records about Betty. By pushing a button, Nancy can access records created by six doctors, eleven nurses, and two clerical workers involved in Betty's check-in, and that's just within New York Central Hospital.

While easy access to such information saves time and untold number of lives, it also creates confusion and pulls medical staff away from patients and onto screens. Medical error is now the third highest cause of death in the United States (Institute of Medicine 2000). No wonder Nurse Nancy lost sight of human connection and of her desire to care for others.

So, what did we witness in the interaction between Betty and Nurse Nancy? We saw the side effects of Digital Technology. As we apply D-Tech, we often make our world more complex. As a result, we frequently lose the ability to connect on a human level, we miss critical insight, and our attempts to coordinate slip into disarray.

Drowning in Potential explores such unintended consequences. Despite the potential embodied in Digital Technology and seen in its direct impacts, the side effects are just as impactful and generally destructive.

D-Tech's direct potential inspires awe: the internet, cell phones, artificial intelligence, blockchain, big data—even traditional spreadsheets and word processors—are amazing tools. Every day brings a new breakthrough that could be used to solve one of society's biggest challenges:

- Game-changing cameras and earpieces can read documents to the visually impaired (Spera 2017).
- Headsets can control prosthetic limbs based on brainwave messages, with the prosthesis providing feedback and sensation (Prattichizzo et al. 2018).
- Virtual and augmented reality can immerse students in learning (Schmitt 2018).

"We're entering . . . the age of abundance," says Eric Schmidt, former executive chairman of Alphabet, Inc., Google's parent company (Brynjolfsson, Rock, and Syverson 2017). We could use D-Tech to solve all—or most—of society's problems. Nonetheless, we all have a Betty in our lives—someone who has experienced a failure of American society due to Digital Technology.

As D-Tech creates new potential, it creates new hurdles, as well. The waves of information and confusion emanating from D-Tech may be figurative rather than literal. But our methods for working together—methods that served us well before Digital Technology—are now drowning in those waves. We are *Drowning in Potential.*

As we'll discuss throughout this book, problems across society remain unsolved. We are barely inching forward—and, in many respects, we're regressing. Type the words, "Society is," into Google, and you'll see that the most common searches characterize society as:

DOOMED	**A LIE**	**BROKEN**
BRAINWASHED	**SICK**	**FALLING APART**

Problems are growing across society's sub-systems of Culture, Government, and Business/Economy.

Our Culture is currently characterized by deaths of despair: the population of Americans without a single friend in whom they could confide grew by more than a third between 1984 and 2004. By 2004, more than half of all Americans (53%) felt that they could only confide in a family member; a quarter had no one—not even a family member to confide in (McPherson, Smith-Lovin, and Brashears 2006). Since 1999, suicides by middle-aged men have increased 43% and women by more than 60% (Curtin, Warner, and Hedegaard. 2016).

We are applying technology to connect with more people in more ways than ever before; social media and video-calling apps put us in contact with loved ones and strangers across the world. Yet, rather than applying technology to bring us together, our culture guides us to develop hollow relationships. Betty's grandson doesn't use his phone to call Betty. And when he occasionally visits her, his eyes and fingers are glued to the digital device.

Our Government is failing as well. Due to inadequate government regulations, the average professional, without realizing it, commits several federal crimes daily (Silverglate 2011, xxv). For example, businesspeople regularly delete emails and physicians often write prescriptions for off-label pharmaceutical uses. Yet these normal, everyday occurrences can be, and have been, prosecuted as felonies (Silverglate 2011).

Government does not use technology solely to protect us from our most destructive instincts. Instead, due to the growth of D-Tech jobs and the ease of creating new digitally stored regulations, new rules are continually written for our multitude of new professions and their expanding capabilities. The U.S. Federal Government currently enforces more than 250 million constraints on our actions. Many of these constraints are so complex and broadly worded that our everyday actions can be indicted as felonies.

Our Business/Economy is also failing to serve society. More than 60% of our economy is dedicated to industries that fail to deliver our most basic requirements. For example:

- The United States spends more on healthcare than any other nation in the world, yet Americans' lifespans are the shortest of any developed country.
- Through the help of Digital Technology, the average American supermarket is now crammed with an incredible 47,000 processed, bred, genetically modified, and preserved foods. Yet these foods do not nourish us. Even our organic foods are less nutritious than their traditionally farmed counterparts from the 1950s and 1960s (Plumer 2015).

While each failure is unique, Digital Technology's unintended consequences are causing fundamental failures in every industry I reviewed.

As a society, we can't turn off technology. Digital Technology has invaded every aspect of our lives—home, work, school, hospitals. D-Tech is here to stay. We need to come together with our technology turned on.

As a parent, an economist, and a citizen, I feel worried. I will pass this society on to my children. And this society must support me in my old age. How can we manage the unintended consequences of D-Tech? How can we survive?

I believe we can survive—and, in fact, flourish—by changing our approach to collaboration. *Technology* does not solve problems. *People* solve problems. And we solve problems most effectively when we work together. Thus, our collaborative approaches must withstand the rigors of today's D-Tech-driven complexity.

In this book, I present insights, strategies, and tools for eliminating the unintended consequences of Digital Technology. Over the following six chapters, we will explore the following questions:

- How have some societies successfully overcome the unintended consequences of past new technologies, when most have not?
- What are the side effects of Digital Technology?
- How can we redesign our approaches to leadership and collaboration so we can overcome the fallout from Digital Technology?

I promise you valuable insight. You will grasp our problems more deeply, and you will encounter new approaches to resolving society's problems.

Don't miss the opportunity to transform your world. Learn why we are *Drowning in Potential*. Become clear about what is happening to our society. Be the one to meaningfully improve society—while benefiting yourself.

CHAPTER 1
TECHNOLOGY AND SOCIETY:

Successes and Failures in Moving Society Forward

In this Chapter, we will explore the relationship between technology and society. The perspective expressed here builds on my experience as a Ph.D. Fulbright Fellow who studied under Gary Saxonhouse, one of the twentieth century's leading economic historians.

Inspiring power comes with new *technology*: ways to significantly impact our environment. Each new invention, from spear to toilet to cell phone, expands our ability to create the world we want.

Yet, solely improving technology is not enough. Society is an adaptive system that must constantly evolve in the face of changing external constraints, challenges, and opportunities. For the past several thousand years, the patterns of these evolutionary challenges have matched today's challenges: new technology has required new forms of collaboration.

Only a minority of successful societies have correspondingly updated their cultural, government, and economic systems and thrived, however. Most societies have stagnated—despite the potential presented by new technology.

For intuition about the impact of technology on collaboration, imagine you and I are digging trenches using wood-handled shovels. To collaborate, we talk to each other. "You dig the left side of the trench, and I'll dig the right side."

Now suppose that we switch from wood-handled shovels to jackhammers. A *side effect* of using the jackhammer's technology is loud noise; we can't hear each other talking anymore. To collaborate, we need non-verbal communication. The change from verbal to non-verbal communication builds on the collaborative approaches we've already developed—yet remains inherently new. And, we fail unless we make the switch (or, at least, one of us likely winds up with a bloody toe). In other words, our world is dynamic:

1. Technology evolves.
2. Society applies new technology, which changes the environment in which society is operating. New forms of collaboration are required for success in the new environment.
3. Society evolves. A minority of societies transform their cultural, government, and business/economic systems to match this new environment, and they thrive. Most societies will stagnate or even regress—despite the potential presented by the new technology.

We've participated in society all our lives. We understand that it is complex (it involves people) and that technology can transform it. At the very least we understand that digital technology has transformed things. We think we must have our cell phones when we leave home or we feel guilty. Mentally, we are "in our phones" instead of with our colleagues at lunch, and we're texting our friends or family rather than talking to coworkers before a meeting. We see evidence all around us of our society being an adaptive system, constantly adjusting to changes in our environment.

I focus on general characteristics of this adaptive pattern in this chapter—the emergence of change, as a systems' scientist would describe it. Figure 1.1 outlines this pattern, using the Agricultural Revolution to illustrate.

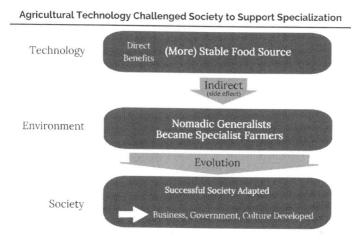

Figure 1.1: As people applied Agricultural Technology, their working and living patterns changed. Successful societies adapted by developing sophisticated business, government, and cultures.

The process begins with society developing and then applying a technology. The top of Figure 1.1 identifies the *direct* impact of such new technology: farm tools feed us, medical tools fix us, and apparel machinery clothe us, for example.

Society also feels *side effects* (i.e., emergent impacts) from the use of such technology as well. When society uses technology, the environment in which we live—which is the environment in which society operates—changes. These changes in our environment, in turn, impact how our society operates. Arrows from the top down in Figure 1.1 illustrate these side effects. For example, as we developed agricultural technology 10,000 years ago, we increasingly chose to settle on farms. Because of this, our entire lifestyle changed. We were no longer nomads chasing the opportunity of the day. We needed to stay home and tend a single crop on the farm for months at a time. Only those societies—a minority—that successfully adapted to this new environment succeeded.

This chapter is divided into two sections. In the first section, we discuss technology and how it develops. In the second section, we discuss how society is impacted by such technology. Throughout this discussion, we define a *technology* as a particular capability to have a significant impact on our environment. Technology is using a steel axe to split a piece of wood. It is using a fishing pole and line to catch fish, and a computer to write a book. Technology is society's fuel; it determines what society has the potential to do.

As we introduced previously, society is a complex, social machine—a system. This system roars to life when we are born. Suddenly, we have parents, doctors, nurses, and even Social Security recorders scurrying around us. Society continues interacting with us every day of our lives, helping us thrive. Society is with us even in death, helping our loved ones mourn and execute our last wishes.

I judge society by the wants and needs its citizens are able to satisfy as they work within the society's culture, government, and business/economic systems. A society that provides for more of its citizens' physical, psychological, and self-fulfillment needs is more successful than one that provides less.[1]

HOW DOES TECHNOLOGY EVOLVE?

As we will discuss in this section, technology evolves like humans, ferns, and mushrooms (see, for example, Kelly 2005; Arthur 2009). It progresses through a process of incremental improvement and a combination of options. And, most spectacularly, this evolution accelerates over time.[2] Evolution follows an intuitive pattern. For example, survival of a pack of wild dogs may depend on their speed and eyesight. Of 100 puppies, the 25 with the best eyesight and the 25 fastest will survive. These surviving dogs will then breed, and some will have the combination of both exceptional sight and speed. Over time, the entire pack improves: it evolves.

How does animal evolution relate to technology? Consider the most awe-inspiring technological feat of the past 100 years:

A young boy sits on the couch next to his dad. His mom sits in a chair a few feet away.

It's July 20, 1969, early evening. Neil Armstrong's voice is coming through the TV speaker... from the moon!

This scene, perhaps more than any other, embodies technological progress. As a nation, we dreamed of that image from the moon. We spent roughly $200 billion to develop the technology to get Neil Armstrong safely to the moon and back (Anthony 2014).

Many who watched Neil Armstrong land on the moon still remember that day. We had succeeded in sending a man from our community to a celestial location far from here. That technology made humankind *believe* that we could do almost anything.

Figure 1.2: This most awe-inspiring technological
feat of the past 100 years was achieved by
combining existing technologies.

However, the moon landing was only spectacular in how it combined existing technologies. Every individual element of the technology used in the moon landing was pre-existing:

- Jet engines had existed since 1930 (Gavrieli, Salim, and Yañez 2004).
- Radios had existed since 1895 (Bellis 2018).
- The vacuum, critical to keep air in a spacesuit, had existed since 1850 ("Vacuum" 2018).
- Even the integrated circuit had existed since 1959 ("Invention of the Integrated Circuit" 2017).

Certainly, the scientists and engineers who landed a man on the moon incrementally improved numerous technologies: they created better jet engines and radios, for example. Few, if any, of these technologies were 100% new to the world though. What was 100% new was a technology capable of taking a man from earth to the moon. Landing that man on the moon. Bringing him safely home. It was the completely new combination of existing technologies that amazed us.

In fact, the moon landing followed the usual pattern. The bulk of technological development is evolutionary; we improve existing elements and combine them into entirely new technologies.

How about vaccines (Dunkelman 2014, 2626)? There's a myth that Louis Pasteur had a brilliant "Aha" moment as illustrated in Figure 1.3. Pasteur injected some chickens with an old, damaged batch of the disease, cholera. When the chickens didn't die, he injected these chickens and others with a new batch of the disease.

The chickens on the left side of Figure 1.3 had missed the first injection and died. The chickens on the right side of the figure had both injections and did not die. The myth is that this one observation was enough to provide Pasteur, the brilliant scientist, with the insight for vaccination: inject a healthy person with a less potent form of cholera (or any other disease) and that person will become immune. Millions of lives will be saved. Brilliant!

But wait! There's more to the story. Louis Pasteur was combining only ideas that had existed long before he was born. Centuries before Pasteur, it was widely known that humans could be infected by smallpox one time only. George Washington knew to "variolate" his soldiers. George Washington's men became immune to smallpox when they were injected with the pus from a smallpox victim! The drawback: 2 to 3% of people variolated died from the purposeful pus injections because they got full-blown cholera. Variolation is dangerous and thus inferior to vaccination.

Also well before Pasteur, Edward Jenner learned that having a mild disease could protect someone from a serious, related disease.

A dairymaid had told Jenner that people exposed to the mild disease of cowpox were safe from the deadly smallpox. The dairymaid had observed this relationship dozens of times in her daily life (dairymaids frequently got cowpox but rarely smallpox); Jenner just added the white coat and the confirming tests to the hypothesis and was rewarded for his insight.

Washington's variolation and Jenner's concepts stood on the shoulders of knowledge and technologies developed by dairymaids and other commoners. Yes, Pasteur's insight that these two technologies could be combined to become a life-saving vaccination was amazing! Nonetheless, that technological insight was evolutionary as much as it was revolutionary.

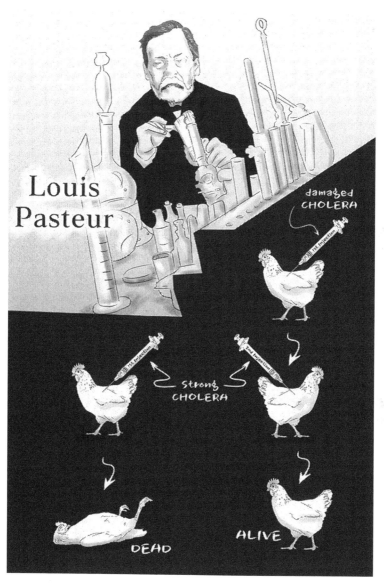

Figure 1.3: Louis Pasteur with his chicken test subjects.
Those chickens who received lethal cholera after
being exposed to an old cholera batch survived. Those
chickens only receiving the lethal disease died.

Most technology improvement is evolutionary, combining or incrementally improving existing technologies. A ballpoint pen is a plastic tube, ink, and a bearing. A pen you use at a bank is a ballpoint pen plus a chain. A printer, at least in concept, is a pen plus a mechanical arm. Similarly, a car is a wagon plus an engine, and a blockchain is a computer plus cryptographic methods. All of the elements of each of these inventions existed before—standing alone. Combined, however, they formed powerful, new technologies.

Evolutionary improvement—including technology improvement—is also predictable (Farmer and Lafond 2016). It improves at a constant rate. Perhaps the best-known example of constantly improving technology is Moore's Law: Intel's ex-CEO, Gordon Moore, famously predicted that the number of integrated circuits on a computer chip would double every two years ("Moore's Law" 2018).

A more accurate description of Moore's law is that the cost of processing information is falling 40% per year (Kurzweil 2001).[3] Moore's law is no longer explicitly holding. It's not possible to place circuits closer together because the distance between integrated circuits can now be measured in numbers of atoms. Yet, manufacturers are now creating semiconductors that can be run in parallel (at the same time). In the future, if we figure out how to operate quantum computers, information processing may operate using a completely new approach. Withal, the underlying rate of processing cost improvement is likely to remain reasonably constant (Farmer and Lafond 2016).[4]

Consistent improvement exists in more than just information processing. Most technologies provide such predictable, consistent improvement rates. Farmer and Lafond (2016), for example, found that all 53 technologies they tested improved at a constant rate. Each technology has its own rate,[5] but that rate appears consistent. For example:

- Dairy milk production per cow increases 2%/year.
- Styrene production costs falls 7%/year.
- Wind turbine costs fall 4%/year.
- Free-standing gas range costs fall 1%/year.

Ray Kurzweil (2005a) illustrates such constant growth in information processing in Figure 1.4, integrating animal evolution and technology:

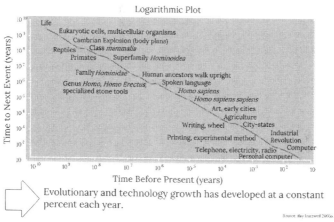

Evolutionary and Technological Growth
"Countdown to Singularity" (Ray Kurzweil)

Evolutionary and technology growth has developed at a constant percent each year.

Source: Ray Kurzweil 2005a

Figure 1.4: Evolution of living things and their machines has advanced information processing at a consistent rate.

Since the beginning of life on earth, evolutionary and technological growth (information processing) have improved a reasonably consistent percent each year. Figure 1.4 can be deceiving because a constant growth rate means an exponentially exploding capability, as in Figure 1.5 (also from Ray Kurzweil [2005b]):

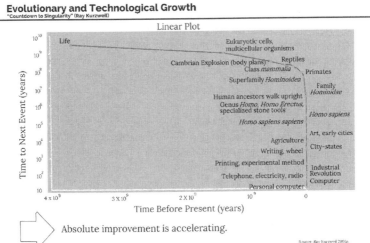

Figure 1.5: Evolution of living things and their machines
has increased the absolute volume of information
that can be processed at an exploding rate.

A constant percentage improvement implies the absolute volume of information that can be processed is increasing at an exploding rate. We're improving faster and faster![6] Technology improvement accelerates because the pool of existing technology grows. The more parts involved, the more things we can improve. We are simultaneously improving thousands of aspects of a car at once, for example. We are redesigning the engine to power the car more efficiently, experimenting with new tire materials to improve traction, and adjusting the angle of airbag deployment to improve safety. As we'll discuss in Chapter 2, we have more than 10 times more parts in an automobile today than were in the Model T—and thus ten times the number of options for improvement.[7]

The more technologies we create, the more options we have for combinations as well. Today we are able to combine any number of billions of existing technologies into even newer technologies. A color television, for example, has morphed from being just a screen in the living room to being an integral part of our phones, the screens in the backs of our minivans, and special LED screens at the gas station.

Yes, perhaps putting a television in the back of a minivan is less exciting than combining a rocket and a radio to go to the moon. However, when we test more and more combinations, we will eventually find our next moonshot.

What has been the impact of recently accelerating technology? The home computer (Apple II) was first sold in 1977 ("Apple II Series" 2018). The Internet was made available to the public in 1991 (Bryant 2011), and the iPhone was first available in 2007 ("History of iPhone" 2018). Technology is evolving—and accelerating.

In summary, technology determines our ability to mold our environment. Over time, as our store of existing technology has increased, so too has the pace of evolution of new technology; the more technology we create, the more tools we combine into new technologies, and the more dimensions for improving existing technologies.[8]

While technology accelerates, however, most societies stagnate. What is happening? Why don't we consistently see the benefits of technology in society's outcomes? That is the topic of the next section.

HOW DOES SOCIETY RESPOND TO EVOLVING TECHNOLOGY?

Society struggles with new technology. As we'll discuss in this section, we are generally capable of managing technology's *direct* impacts. Too frequently, however, we adjust ineffectively to technology's *indirect* impacts—its *side effects*. Like the need to switch to non-verbal communication when upgrading from the quiet hand-shovel to the loud jackhammer, new technology requires society to implement new modes of collaboration.

Bright and shiny mesmerizes us. We're hypnotized by the glittery, new technologies that come out of our laboratories and universities:

- A massive robot that can smash another robot!
- A currency that exists just in cyberspace, which the Government cannot control (bitcoin.org)!

We are inspired by the potential to do things more quickly and easily, saving money in the process. We are excited by dreams of improving quality and expanding variety. We love to look at all of the amazing, positive things technology can directly do for us, such as the ideas highlighted by Figure 1.6, based on an image by Dr. Stephen Omohundro (2017).

On the other hand, evil gives us nightmares. Nobel Prize winner Muhammad Yunus explains, "While technology is important, it's what we do with it that truly matters (2012)." We fear that the self-flying drone will be fitted with explosives or the computer will be used to hack into our bank accounts. These concerns, such as in Figure 1.7 below, also based on an image by Stephen Omohundro, are certainly the focus of many people's thoughts today—and we must manage such concerns.

And there's more. As we'll discuss over the rest of this chapter, a focus solely on what technology directly delivers is too narrow. We miss technology's broader impacts, technology's *side effects*. Professors Kate Crawford and Ryan Calo (2016) call this omission of technology's side effects a "blind spot" in technology evaluation. As we'll discuss further, technology changes our environment—and, when this impact is strong, society must actively transform itself—we must find new ways of working together. The *side effects* of applying powerful technology have made people in some societies *worse off* (or even cause some to fail altogether).

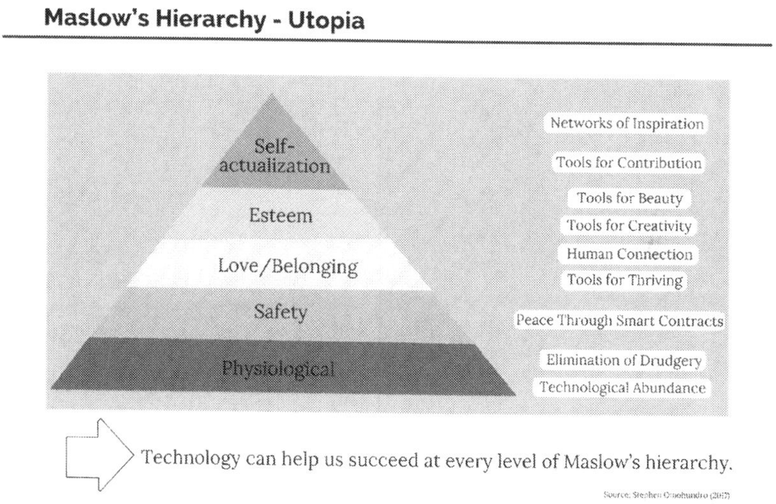

Maslow's Hierarchy - Utopia

Technology can help us succeed at every level of Maslow's hierarchy.

Source: Stephen Oonobundo (2017)

Figure 1.6: By applying Digital Technology, we could deliver benefits at every level of Maslow's Hierarchy.

How does technology impact society? The same way pharmaceuticals impact the human body. We've seen the commercials. Attractive actors swim across the screen while the announcer describes the "75% and even 90% elimination" of some ailment. "[Pharmaceutical X] works inside the body to help block a specific source of inflammation that can trigger [a common skin condition] (Humira 2017)."

The pharmaceutical changes how our bodies operate—which can indirectly cause side effects throughout that human system. Only if the body's parts appropriately adapt their mode of working in a coordinated manner will side effects be avoided. As the announcer in the commercial explains, "[Pharmaceutical X] can lower your ability to fight infections including tuberculosis. Serious and sometimes fatal infections and cancers . . . have happened . . . as have blood, liver, and nervous system problems."

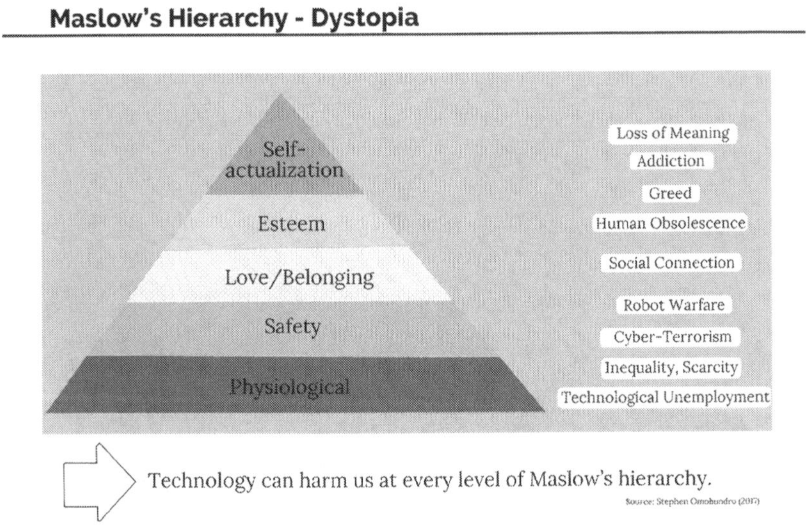

Maslow's Hierarchy - Dystopia

Self-actualization

Esteem

Love/Belonging

Safety

Physiological

Loss of Meaning

Addiction

Greed

Human Obsolescence

Social Connection

Robot Warfare

Cyber-Terrorism

Inequality, Scarcity

Technological Unemployment

Technology can harm us at every level of Maslow's hierarchy.

Source: Stephen Omohundro (2017)

Figure 1.7: By applying Digital Technology, we could deliver harm that impacts us at every level of Maslow's Hierarchy.

Can a pharmaceutical targeted at our skin really impact our nervous system? Yes. In such a system as complex as the human body, once you change how one aspect of the system operates, you can indirectly cause *side effects* that may emerge in any element of that system. The same is true in a system as complex as modern society.

In this section we discuss the challenges societies face as they adapt to new technology. We begin by looking at societies that successfully adapted during the Agricultural Revolution. We follow by discussing societies that failed to adapt during the Agricultural Revolution and at other times.

SUCCESSFUL ADAPTATION

Ten thousand years ago the Agricultural Revolution transformed society. We began to farm: scratch the earth, drop in a few seeds of grain, and hope those seeds grow. It may not sound like much. However, this technology fundamentally changed society.

Successful societies replaced their pre-Agricultural Revolution structure which featured tribes of nomadic generalists with markets filled with specialized settlers. Such specialists needed business to facilitate trade, government to support that trade, and increasingly developed culture to nourish specialist knowledge.

Before Agricultural Technology, collaboration was straightforward. "Bjorn and Mac go hunt while Ulna and Apathena gather berries. Tomorrow, Bjorn and Mac will carry our belongings while Ulna and Apathena will herd the children." Any man could do any man's task, while any woman could do any woman's task (Ember 2014).

Successful societies restructured themselves after they developed agricultural technology. Peter Drucker (1970) describes the Agricultural Revolution as: "The first great revolution technology wrought in human life."[9] This revolution began humbly, however, as Yuval Harari (2015, 174) describes:

> In the years following 9500 BC, the descendants of the Natufians continued to gather and process cereals, but they also began to cultivate [cereals] in more and more elaborate ways. When gathering wild grains, [the Natufians] took care to lay aside part of the harvest to sow the fields the next season. Gradually, they also started to weed the fields, to guard [their fields] against parasites, and to [fertilize] and water them.

Like a pebble dropped into the water, irrigating those fields, the impact of agricultural technology rippled across society. As more time was spent farming, there was less time to gather and hunt wild species. "The foragers became farmers (Harari 2015, 175)." They settled on their land.

We can imagine the resulting change in society's environment. Nomads turned peasant farmers looked like Horacio and Daksha (based on Harari 2015, 176, 200):

> *Horacio looks down at his little plot of land. He sees the store of grain piled neatly by his hut and he looks up at his wife, Daksha, with her gnarled fingers. He smiles. "Yes. I am lucky," he thinks. No longer will he and Daksha wander over dozens of acres of land chasing the opportunity of the day; they will stay at home, creating riches with their farmland.*

> *Life is bittersweet. Daksha has survived being pregnant every year for more than a decade—a feat Daksha and Horacio's nomadic ancestors could never have matched. Seven of their eleven children who were born, survived. Five of their children are strong and the other two are survivors.*

> *With so many children comes the promise of help in the fields, making Horacio smile. Four of the children are helping in the field already and the other three will be soon. They will all learn the intricacies of successful farming.*

> *The nature of Horacio's work has challenged him to specialize. His nomadic ancestors faced a new challenge each day where failure one day could be made up by success the next. Did he miss the antelope with his spear-throw today? Perhaps rabbits will be better prey tomorrow.*

> *On the other hand, after spending months tending a single crop, Horacio must be successful virtually every season if his family is to prosper. Yet, successful farming requires understanding sophisticated nuances about land, weather, and the crops being grown. Horacio has had to learn to plant on rolling*

hillsides and to recognize the crops that grow best in the shade of his particular plot of land. In other words, Horacio has had to become a professional.

Daksha also has specialized. She now makes ceramic plates. Other families have settled close by as this region has the most fertile land, and many are able to produce more than enough to just feed themselves. Many want to use that surplus for luxuries beyond food, such as Daksha's plates. Daksha has identified the best types of clay, identified a process to ensure consistent quality, and designed a glowing glaze finish.

Daksha and Horacio are not alone in such specialization. Harari state that over time, increasingly populated regions provided full-time employment for professional shoemakers, doctors, potters, and carpenters (Harari 2015, 209).

The challenge for Agricultural Revolution societies was creating the systems whereby such specialists could effectively collaborate. The core of such collaboration was trade, for which the tools of business and the economy were critical. During the Agricultural Revolution successful societies developed the core tools that economists, even today, recognize as the heart of competitive markets.

Price played a key role in such societies, as it effectively told even large numbers of specialists what to do more and what to do less. In 1945, Friedrich Hayek (1948, 87) described the power of price to coordinate activities across a society of specialists:

[Price operates] without an order being issued, without more than perhaps a handful of people knowing the cause, [as many as] tens of thousands of people whose identity could not be ascertained by months of investigation, are made to [respond.] When the price of a scarce resource such as rice rises, for example, people are inspired to [increase production of that grain, and others to] use that grain more sparingly; that is, they move in the right direction [to balance supply and demand].

Figure 1.8: Agricultural Revolution families were settlers, not nomads, and they became the first specialized professionals.

Horacio, himself, monitors the price of rice and sorghum. When the price of rice is higher, he plants more rice. When the price of sorghum is higher, he plants more sorghum. The high price signals a shortage, and Horacio's self-interested actions help reduce that shortage.

With the principle of price has come the introduction of money, which is the language of price. In wicker baskets, Horacio and Daksha store cowry shells that can be used for trade. Horacio and Daksha cannot understand the full breadth of their usefulness, but such shells can be used as money as far and wide as Africa, South Asia, East Asia, and Oceania (Harari 2015, 351).

With money, citizens during the Agricultural Revolution took their Business skills to the next level through borrowing and lending (Harari 2015, 110–11). When societies with agricultural technology developed such Business skills, their citizens became physiologically better off.

In the most successful societies, more than just Business evolved. Government, also, required upgrading for effective collaboration in Agricultural Revolution society.

Consider what happened in Horacio and Daksha's community. For many years, Horacio and Daksha reaped surplus harvests, and such surplus attracted other settlers. As Harari (2015, 209) describes, "Around 8500 BC the largest settlements in the world were villages such as Jericho, which contained a few hundred individuals . . . By the fifth and fourth millennia BC, cities with tens of thousands of inhabitants sprouted in the Fertile Crescent."

Horacio wakes from his reverie about his life with Daksha. A man, Lorenzo, walks toward Horacio from across the village. "Horacio, the Hessian flies ate much of my wheat crop this year. It looks like you have done well. Perhaps you could pay me to make you some ale." Lorenzo carries a spear but doesn't look threatening. However, Horacio doesn't know Lorenzo well. Can he trust Lorenzo?

Horacio can trust Lorenzo if a competent Government supports contractual agreement between the two and sets the limits of acceptable behavior. As Peter Drucker describes, the most successful Agricultural Revolution Business "required [development of] the first . . . impersonal, abstract, codified legal system." This first great code of law of a most four thousand years ago, the Code of Hammurabi,covered, for example ("Code of Hammurabi" ca. 1780 B.C.E.):

- *Employment Law (Rule 265): employees shall not defraud their employer*: "If a herdsman, to whose care cattle or sheep have been entrusted, be guilty of fraud and make false returns of the natural increase, or sell them for money, then shall he be convicted and pay the owner ten times the loss."

- *Commercial Law (Rule 104):* transactions require transfer of goods, money, and receipts: "If a merchant gives an agent corn, wool, oil, or any other goods to transport, the agent shall give a receipt for the amount, and compensate the merchant therefore. Then he shall obtain a receipt from the merchant for the money that he gives the merchant."

- *Liability (Rule 53):* If your lack of effort causes damage, then you must pay a penalty: "If any one be too lazy to keep his dam in proper condition, and does not so keep it; if then the dam breaks and all the fields be flooded, then shall he in whose dam the break occurred be sold for money, and the money shall replace the corn which he has caused to be ruined."

We might consider Hammurabi's methods of enforcement coarse. Yes, rule 53 says that if you let your dam break then you will be sold as a slave. And Hammurabi's Code is famous for Rule 196: If a man puts out the eye of another man, his eye shall be put out (an eye for an eye). Nonetheless, "the first great code of law (Harari 2015, 110–11)" would still be applicable to a good deal of legal business in today's society.

Such a legal system required the first permanent Government.[10] This permanent Government hired a staff—a new set of specialists. That staff of specialists had a clear, hierarchical structure, and "soon there arose a genuine bureaucracy, which allowed irrigation cities to become irrigation empires (Harari 2015, 109)." As Harari (2015, 209) describes:

> In 3100 BC the entire lower Nile Valley was united into the first Egyptian kingdom. Its pharaohs ruled thousands of square kilometers and hundreds of thousands of people. Between 1000 BC and 500 BC, the first mega-empires appeared in the Middle East: the Late Assyrian Empire, the Babylonian Empire, and the Persian Empire. They ruled over many millions of subjects and commanded tens of thousands of soldiers.
>
> In 221 BC the Qin dynasty united China, and shortly afterwards Rome united the Mediterranean basin.

When societies with agricultural technology developed such Government, citizens were safer and trust was higher. Yet, to most effectively collaborate in a society with Agricultural Technology, more than just Business and Government developed. Culture, also, had to evolve.

Horacio embodies such an evolved culture. Despite the hardship, Horacio feels pride.

> Horacio watches in the evening as his son, Quance, studies his lesson from school. Quance can read and write and has begun to record the family's grain inventory.
>
> The hut and its surplus grain belong to Horacio and Daksha, not Lorenzo. If Horacio works hard, he will be rewarded. Government and Business actions are focused on the

individual. The individual's relationship with the broader group is secondary.

This order of importance reversed the relationship that existed before the Agricultural Revolution in many societies. Before this revolution, a person was frequently a member of a tribe, first and foremost, with the individual secondary. However, "in the [agricultural] city of antiquity . . . the individual became, of necessity, the focal point (Harari 2015, 110–11)." They became a citizen of the agricultural city.[11]

Also during the Agricultural Revolution, writing developed as a Cultural tool to support specialists in the most successful societies. Specialist businesspeople "needed records, and this, of course, meant writing (Harari 2015, 111)." Specialist Government required writing of rules and regulations for all to see.

While other societies struggled to grow, the Sumerian society developed such a writing capability, and that society's success grew rapidly (Harari 2015, 2011). Legal systems and business records of all sorts developed and were written on parchment, clay tablets, and stone, as a result (Harari 2015, 111).

Writing brought with it more sophisticated social discourse, which brought people together. As Harari (2015, 209) describes, "When the Agricultural Revolution opened opportunities for the creation of crowded cities and mighty empires, people [in successful societies] invented stories about great gods, motherlands, and joint stock companies to provide the needed social links . . . The human imagination [built] astounding networks of mass cooperation, unlike any other ever seen on earth."

Collaboration was easier with such stories. Religious stories, for example, defined a shared purpose for everyone's activity. Such shared purpose allowed individuals to focus on how to cooperate rather than on answering why. The stories also clarified more tactical expectations of each other. The Bible, for example, "in addition to being the divine word of God that would guide people through life's journey to the next world, served as a textbook for history, a source book for morals, a primer for mothers to teach their children how to read, and a window through which to view and understand human nature (Kaminski 2002, 1)." This quote was written to describe the impact of religion during the American Civil War. However, such words equally describe religion's role of several thousand years earlier, as well.

In summary, successful Agricultural Revolution societies had to develop new modes of collaboration. Farmers could no longer maintain the lifestyle of generalist nomads; they had to specialize. Such specialists collaborated through Business development, which provided physiological benefits to citizens. To collaborate as specialists through Business, professional Government was required, which enhanced safety and trust. And, Business and Professional Government required an enhanced Culture, which advanced social cohesion and nourished self-actualization.

If that were the end of our story, it would be a beautiful story. Yet, our story is not complete. Not all societies evolved into such wondrous successes during the Agricultural Revolution. Many did not. In fact, most societies stagnated or worse, despite the advent of agricultural technology. Such failure is the focus of our next section.

FAILURES TO ADAPT

Yuval Harari (2015, 176) suggests that every Agricultural Revolution society failed:

*With time, the 'wheat bargain' became more and more bur-
densome. Children died in droves, and adults ate bread by the
sweat of their brows. The average person in Jericho of 8500 BC
lived a harder life than the average person in Jericho of 9500
BC or 13,000 BC. But nobody realized what was happening.
Every generation continued to live like the previous generation,
making only small improvements here and there in the way
things were done. Paradoxically, a series of 'improvements,'
each of which was meant to make life easier, added up to
a millstone around the necks of these farmers . . . Farming
enabled populations to increase so radically and rapidly that no
complex agricultural society could ever again sustain itself if it
returned to hunting and gathering.*

Harari's premise may be extreme. Yet, Harari's main point remains valid with even a moderate view of society: many societies
failed to provide their citizens the benefits agricultural technology promised. Not every society constructively responded
to the impact of new technology. They allowed themselves to
maintain Business, Government, and Cultural systems that did
not facilitate the type of collaboration agricultural technology
required. Some societies were immediately overrun by marau-
ders, or simply disappeared. Others never went on to develop
philosophy, poetry, music, and art. Each of these societies failed
to one degree or another.

Peter Drucker (1993) describes how Native American societies created the root of their own failure. In Native American
societies:

*The individual, for instance, failed to make his appearance.
Never, as far as we know, did these civilizations get around to
separating law from custom nor, despite a highly developed
trade, did they invent money.*

Without such critical transformations, Native American societies were primed for failure. Citizens of these societies were unable to collaborate as specialists to the same degree as societies that did develop the individual, law, and money, such as those in Europe. As a result, Native American societies failed to propel forward to the same degree by agricultural technology. Instead, they stagnated while European societies continued to progress.

By the time the white man came from Europe to America, Native American societies had a fraction of the competitive capabilities that the Europeans did. In many cases, it was the diseases the white man brought from Europe that wiped out the Native Americans. Yet, even had the Native Americans avoided disease, their society would have been primed for domination by the Europeans.

The key to dominance was superior societal systems: Culture, Government, and Business sub-systems that best support collaboration in an environment featuring the day's technology. Societies more skilled at developing technology were routinely dominated (Harari 2015, 515). The Sassanids, for example, had superior military and civilian technology relative to the Arabs—but were defeated. The Byzantines had better technology than the Seljuks, and the Chinese possessed better tehnology than the Mongols. In all cases, the better technology lost out to the superior societal systems.

It may seem like we live in a different world today. Consider Figure 1.9, which illustrates economic development for global regions since 1600. Economic development is measured in that chart by the value of goods and services created by each person in a country (percapita GDP, adjusted for inflation and exchange rate). Regions with the most advanced technology, such as those in North America, Europe, and Japan also have experienced the greatest economic development over this period. These are the regions with lines at the top of the chart.

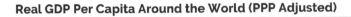

Real GDP Per Capita Around the World (PPP Adjusted)

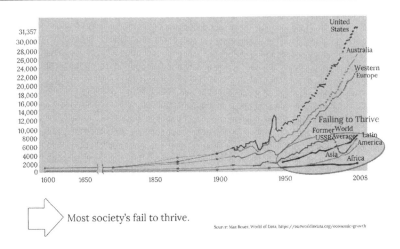

Figure 1.9: A few countries (such as the U.S.) have achieved far greater economic growth than all others, for hundreds of years (Roser 2016).

Yet, consider more carefully. If you are lounging on your sofa in Kathmandu, could you obtain access to cutting edge technology? You could. Simply turn on your computer and search the Internet for courses, perhaps, on neural networks. Courses covering everything from fundamentals through the most sophisticated doctorate level are at your fingertips. Purely technological innovations are easy to duplicate—anywhere in the world.

What is not available in Kathmandu are the Cultural, Government, and Business systems that have served the Americans, Europeans, and Japanese so well throughout their Industrial Revolutions.

Today, at least 80% of humanity lives in the regions represented on the bottom of Figure 1.9—with most of those people living on less than $10 a day (Shah 2013). These people live around the world, from Latin America, to Western Asia, to Eastern Europe, Africa, and more. For hundreds of years, societies in these less developed countries have largely stagnated.

For those hundreds of years, the people in less developed nations have had access to much the same technology as is available in America. Yet, these societies have failed to correspondingly benefit. Their societies failed to develop the Culture, Government, and Business systems that would support the collaboration needed to benefit most from industrial technology.

The primary relationship is not from technological development to societal benefit. Critical is development of the societal sub-systems—Culture, Government, and Business—that will support the type of collaboration needed to leverage the day's technology. Societies able to support that type of collaboration progress while those that are unable, stagnate or regress.

SUMMARY: THE RELATIONSHIP BETWEEN TECHNOLOGY AND SOCIETY

Technology is closely related to a society's success. However, this relationship is not direct. Citizens do not necessarily benefit when a wonderful technology is available.

In many respects, technological development can be considered separate from societal growth. Technology evolves based on the volume of existing technologies that can be combined. As more technology is developed, technological development accelerates.

Societies flourish when they develop the societal sub-systems (Culture, Government, and Business) that allow citizens to collaborate in a world featuring the new technology. A minority of societies successfully responds to each new generation of technology so that the citizens of those societies benefit. Most societies stagnate.

Over the Agricultural Revolution, successful societies found new ways to support collaboration among specialists. Such societies supported Business's market mechanisms, professional Government, and a Culture emphasizing the individual. The core of the Culture, Government, and Business systems of even today remain the same.

In other words, for the past 10,000 years, our most successful societies have continued to support specialization. We've developed wave after wave of new technologies with the most recent one being Digital Technology. What new hurdles to effective collaboration does D-Tech pose? Is the U.S. developing the societal sub-systems necessary to translate that technology into attractive societal outcomes? Or, are we on the path to becoming a nation that stagnates in the face of powerful technology?

Let's take the next step toward answering those questions by looking at D-Tech: what is it and what are the unique challenges to collaboration when applying it? Answering these questions is the focus of Chapter 2.

CHAPTER 2

DIGITAL TECHNOLOGY:

*Potential and the 3 Ways It's Making
Our Lives More Challenging*

In this chapter, we look at Digital Technology (D-Tech) and examine the impact of that class of technologies on our environment. D-Tech includes any technology that digitally processes information: computers, smart phones, software, artificial intelligence, blockchain, and more. This chapter builds on my years learning about what D-Tech is and how it works as a partner and sounding board to one of Silicon Valley's pioneers in artificial intelligence safety, Dr. Stephen Omohundro.

Today, we need to invent modes of collaboration that complement Digital Technology (D-Tech). Digital Technology is like a jackhammer: it provides amazing power, yet also creates *side-effects* impacting how we must collaborate. A systems scientist would call these digital technology impacts *emergent*: arising as a natural or logical consequence of D-Tech, yet not uniquely traceable to that origin.[1]

Figure 2.1 illustrates the process through which D-Tech is impacting our collaboration:

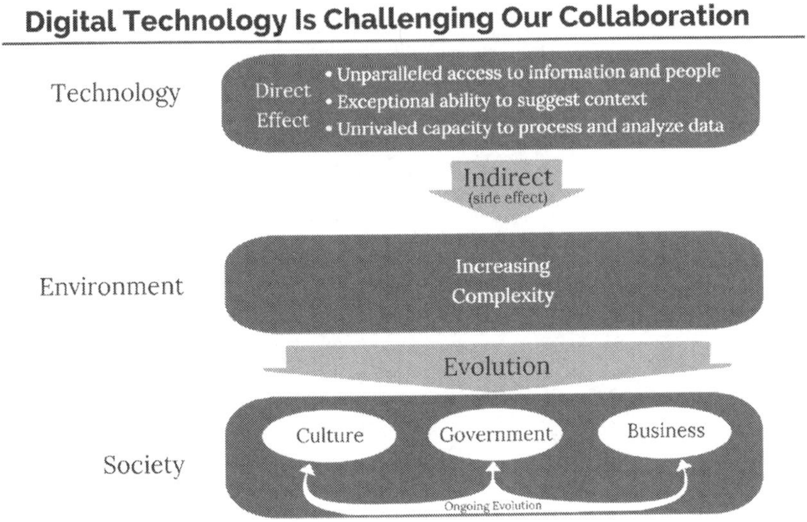

Digital Technology Is Challenging Our Collaboration

Figure 2.1: As we apply Digital Technology, we are creating an ever more complex society. This complexity, in turn, is causing our society to evolve.

D-Tech provides three direct benefits, listed at the top of the figure—each of which creates a *side-effect* of increasing complexity, represented by the middle of the figure.

1. **Unparalleled access to information and people:** Suppose we want to know the characteristics of fatty acid molecule C-16-2. In this situation, digital technology provides us unparalleled access to millions of related information sources at the click of a button. That's powerful.

   We love this D-Tech power. Yet, with the onslaught of digital technology, we've become inundated with content. We now publish 40 times as many books every year as we did in the 1960s (Piersanti, n.d.). We publish two and a half times as many journal articles every year as existed, in total, in 1950. With this much information, we must rely on D-Tech to help us find the content we need.

All too frequently, however, we cannot find the information we need to make good decisions. If I'm trying to decide between eating a chicken pot pie and a roast beef sandwich, for example, I can't find advice about which would be more healthful for me. The millions of web pages about the molecule C-16-2 may be relevant—but they only confuse me in the end.

2. **Exceptional ability to suggest context:** Suppose we want to understand how American business leadership operates.If we look at Google Images, we are likely to find 100 images for CEO: 99 images of male CEOs and 1 of a CEO Barbie doll (Crawford 2017). The content of those images is factually correct: all 100 images are of CEOs. However, that mass of information implies a particular context—that we live in a world in which real CEOs can only be men. Similarly, when our phone *pings*, our brain interprets a particular context—that the email the *ping* announced is currently the most important information in our vicinity. Such implied context has as much influence over our decision-making as the information's actual content.

Unfortunately, this D-Tech-implied context can destructively manipulate our decision-making. Even if only 4% of Fortune 500 CEOs today are female, we may want to actively support a different perception of reality to help empower women (Zarya 2016). The *ping* from our phones may interrupt a meaningful conversation with a loved one, only to focus us on the most recently emailed advertisement. D-Tech's ability to influence our thought is amazing—yet that power is only beneficial for society if the impact is constructive.

3. **Unrivaled capacity to process and analyze data:** D-Tech provides us unrivaled capability to analyze the impact of our decisions. Perhaps we want to know what happens to profit if we reduce inventory by 10%. Push a button, and we have our answer. Ray Kurzweil estimates that one computer will be able to analyze as much information as one billion people by 2040 (Diamandis 2015). That's awe-inspiring.

   We have leveraged this analytical capability to create an amazing diversity of options and specialists. The impacts of our decisions are deeply interrelated with decisions made by many of these specialists. We each oversee only small elements of the large, interconnected systems that deliver what we want and need.

   As a result, collaboration today requires aligning with dozens of types of people. For example, with the masses of specialized technicians, physicians, orderlies, etc. involved in modern healthcare, Nurse Nancy interacts with 130 different types of people (Crooks 2018). An interaction with any one of these types of people can shift the focus of Nancy's all-too-harried days—thus indirectly impacting her interactions with everyone else.

   D-Tech helps us coordinate the most tactical decisions. D-Tech can help ensure that Dr. Rosen and Dr. Chai do not prescribe pharmaceuticals that, in combination, would kill a shared patient. Nonetheless, we struggle to coordinate both the massive number of decisions we face and the more strategic aspects of our decisions. Those two physicians whom technology restrains from killing their shared patient must actually speak with each other if they want to coordinate on treatment philosophy.[6]

We will return to Figure 2.1 in chapters 3, 4, and 5 to discuss the bottom portion—the impact on our society.

Almost 50 years ago the Digital Technology Revolution began to transform society. We started applying D-Tech: gathering data with which to inspire new thoughts and store it in computers, like the seeds for next year's harvest; applying analytical tools that nourish clear thought like the sun, water, and fertilizer nourish plants; and hoping to harvest better ideas and decisions, like we hope to harvest a wealth of food on our farms.

As D-Tech began to be implemented, society began to fundamentally change. Do you remember reversing pay phone charges because you didn't have any change, or traveling with cassette tapes strewn around the car? How about writing a report *without* cut and paste, or adding a column of numbers on a calculator then, re-entering the entire column to check the answer?

Our environment is different because of D-Tech, that's for sure. The impact of D-Tech has rippled across society. As we will see, these ripples have much in common with the impact Agricultural Technology had on society thousands of years ago.

In Chapter One, we discussed how Agricultural Technology, directly, only helps us grow plants in the soil. Nonetheless, Agricultural Technology's impact went far beyond plants; it irrevocably changed society. Most societies stagnated rather than materially benefitting from the promising technology. The minority of successful societies replaced their old structure of nomadic generalists with markets featuring specialized settlers. Such societies developed the increasingly sophisticated cultural, governmental, and business/economic sub-systems required to support collaboration among such specialists.

For thousands of years now, the most successful societies have continued to develop such specialist-supporting sub-systems. As the Digital Technology Revolution gains momentum, specialists from heart surgeons to comedians remain well supported by society's sub-systems.

Society's greatest challenge now lies in supporting us in areas where we are *not* specialists. We have come full circle. No longer are we effective generalists challenged to support specialists. We are now effective specialists challenged to bring our expertise together to support us in areas where we have no special expertise.

We discuss today's challenges first in terms of *Direct Impacts*, followed by the resulting *Indirect Impacts* (also called *side effects* or *emergent impacts*.)

WHAT ARE DIGITAL TECHNOLOGY'S DIRECT IMPACTS?

We begin by discussing the top portion of Figure 2.1: the direct impacts of D-Tech. Digital Technology is a powerful toolbox. Beginning in approximately 1970, the Digital Technology Revolution began to change the way we think.[3]

Our D-Tech world is a specialist's dream. D-Tech is exceptionally powerful at helping us gather detailed information. The information D-Tech selects for us, and when and how it presents that information, influences our perception of our world's limits and how that world works (i.e., our context). And, D-Tech provides increasingly powerful tools to analyze the implications of the options. In sum, D-Tech has a direct effect on our ability to carry out important information-processing tasks: Gathering Information, Understanding Context, and Analyzing Options.

For example, with D-Tech, we can find 4,793 hotels in Budapest (Gathering Information).[4] As we dig into the details about a first hotel, we see a pop-up ad for Prague, convincing us to think more broadly about our possible destination (Understanding Context). Then, with a few mouse clicks, we can compare amenities and prices across hotel rooms, imagining what it would be like to stay in each (Analyzing Options).

UNPARALLELED ACCESS TO INFORMATION

D-Tech provides unparalleled access to information and people. By 1982, D-Tech was able to store and retrieve inventory and other quantitative records. We could search a spreadsheet for the number of size 5 screws in inventory and plan our next order—without ever leaving our desks. At the time, such search capability was striking, and D-Tech today retains that critical ability to search through vast volumes of information and to access any information it can recognize.

But those capabilities were just the start. With the Internet, D-Tech gained the ability to access most digitally stored information from anywhere. We fell in love with this ability, so now we're creating explosive volumes of data. We're putting digital sensors in everything from airplane engines to armchairs, all to feed D-Tech with ever more data. We've labeled this trend the "Internet of Things": every measurement possible, available all the time.

Almost any piece of data a specialist desires is readily available. Want a reference on the impact of fatty acid molecule C-16 on the functioning of the liver? Google provides 4.6 million sources. Want to know the singer Cher's birthdate? Google provides more than 12 million sources.

D-Tech's ability to search out and gather information for us has recently been extended. D-Tech has always been good with precisely defined concepts such as "the number 3." Yet, until recently, D-Tech couldn't gather information about real-world things, because the real world simply doesn't match precise definitions.

Consider, a search for an image that includes a bird. Ask children to describe a bird and they're likely to say something about an animal with feathers, that lays eggs, and sleeps in a nest. However, as Figure 2.2[5] exemplifies, we recognize a bird even if it has none of those simple characteristics. We recognize the "bird-like eyes," a "bird-like bill," and a textured neck scruff that "kinda looks like it could come from feathers."

Figure 2.2: A picture of a goose. Even though the picture doesn't clearly show any feathers, eggs, or that the animal lives in a nest, we clearly recognize it as a bird.

D-Tech can now recognize birds using a similar approach: searching for recognizable characteristics.[6] Real-world implications include that Facebook can send you a picture of your Uncle Murray in a photo despite an uncharacteristic scowl and his odd pizza-man costume. Google's self-driving cars can recognize and respond to other cars in their environment as well as people, birds, and lampposts.

Using similar concepts applied to sound, Apple's Siri and Amazon's Alexa can now recognize and pass on information related to (many of) the words you say—despite your lisp, your rolled r's, or your slurred *th*'s.

Withal, D-Tech does more than just duplicate human abilities to gather and pass on information. D-Tech is something different. With regards to Gathering Information, D-Tech can guarantee that it is not making anything up; when programmed appropriately, D-Tech can guarantee that any information that comes out is as truthful as the information that went in.[7]

Unfortunately, D-Tech's interpretation of the information that "went in" can be influenced. Researchers call it an *adversarial attack* when someone tries to influence D-Tech's perception of the information it receives. For example, the following adversarial attacks are possible (Mikhailov and Trusov 2017):

- Change just a couple of pixels ("add slight visual noise") to a $100 check, and deposit it in your ATM for $1,000,000.
- Add modest visual noise to a speed limit sign so that self-driving cars think they should be targeting a speed of 200 miles per hour.
- Redraw your license plate with some visual noise so that humans can still read it—but D-Tech thinks you have no license plate.

In short, D-Tech is increasingly able to help us in Gathering Information. D-Tech can search billions of records for exactly what you're searching. As D-Tech evolves, the number of things it can recognize and pass on as part of this information gathering is expanding. Yet, this ability to gather information differs from human information gathering. A second set of eyes or ears? How about a million sets—that will tell you only the truth . . . unless someone plays them a garbled recording that only they can interpret?

EXCEPTIONAL ABILITY TO INFLUENCE OUR UNDERSTANDING OF THE WORLD

We rely on D-Tech-driven information for insights into how our world is operating. I call such insights, "understanding context." For example, an Internet search may identify that less than 5% of Fortune 500 CEOs are female. From that, we are likely to interpret (the context): males are more likely to achieve American business leadership success than are females.

At least as influential on our thinking as the data itself is how and when D-Tech provides us that data. "If you do a Google Image search for CEO, you get a lot of white dudes in suits . . . The first female that shows up . . . is CEO Barbie (Crawford 2017)." That long lineup of male images (and one, almost comically, female image) provides us an even more powerful mental image of our world than does the data about CEOs. In the example about finding a hotel in Budapest a few pages ago, D-Tech's well-timed pop-up ad convinced us to broaden our consideration of travel options to include Prague as well as Budapest.

This D-Tech ability to influence our understanding of context is supercharged by our human thought patterns. For example, we share fake news more rapidly than we share the truth online (R. Meyer 2018). We spend time doing things that leave us depressed (Shakya and Christakis 2017). And we consistently make the same errors in our mental math. D-Tech can be programmed to leverage these tendencies to influence our understanding of the world around us—and how we react to it.

Ironically, D-Tech, itself, does not truly understand how our world works. D-Tech can only do one of two things:

1. Provide information that helps us understand context as precisely defined for the D-Tech by a human controller.

2. Reflect data in the machine's environment that may (or may not) suggest the same context a human would select to portray. For instance, we humans are likely to expand our children's thinking by clarifying that we live in a world in which each gender is able to exhibit leadership strengths. Such a perspective is more constructive than one focusing solely on women's currently limited success in achieving business leadership positions. D-Tech is unable to recognize such far-reaching perspectives by itself.

What does D-Tech's lack of understanding context look like? Imagine an automated translation machine interpreting a female (Nadiya) saying, "Jason, it's time," as a comment about the time of day. When Nadiya is standing with you offstage as you prepare to speak, the machine is correct. Yet, we humans understand Nadiya's message to be completely different when you and she, as newlyweds, are alone in a dimly lit bedroom. The machine is unlikely to understand the difference between the situations—interpreting both utterances as a signal to begin a grand speech.

D-Tech's limitations in understanding context are beginning to decrease. Consider, for example, the novel *Dracula*, by Bram Stoker. The main characters, Lucy and Arthur, get married toward the beginning of the book. Within the context of such a loving relationship, we would think Arthur was deranged if he stabbed Lucy.

However, the context of the relationship changes: Lucy becomes sick. She dies and becomes reanimated as a Dracula monster. We understand that the context—the relationship—has changed. As a result, we also understand that Arthur is sane when he stabs Lucy's reincarnated corpse in the heart.

D-Tech can now map the changing relationship between Dracula's Arthur and Lucy. The technology, itself, can now read and label the sections of the book: love, marriage, death, and fighting (Iyyer 2017). Yet, D-Tech did make some nonsense descriptions of Arthur and Lucy's relationship, in addition to those that were spot on.[8] Such D-Tech is far from perfect.

In short, despite recent improvement, D-Tech's role in helping us understand that context is different from that of humans. D-Tech, *itself*, is less skilled than we are at understanding how our world works. Yet, D-Tech can be programmed to influence *our* understanding of context (and how we react to that context) in incredibly powerful ways.

UNRIVALED CAPACITY TO ANALYZE OPTIONS

Once we understand how the world works, we need to compare (i.e., analyze) our options. In seconds, D-Tech can calculate the impact of our reducing inventory by 10%. Even as early as 1985, D-Tech could help us add a massive list of numbers. D-Tech was, even then, massively faster than any human's brute-force during a computational task. Through such information processing, we analyze the data we have until we are confident in the recommendations of whatever analysis we are conducting.

D-Tech's ability to process and analyze information is rapidly growing. So-called machine learning algorithms are increasingly capable of interpreting masses of data ("Choosing the Right Estimator," n.d.). Combined with improved D-Tech hardware, machines can now identify incredibly subtle patterns. Almost instantaneously, the software can identify which among the millions of daily New York Stock Exchange trades are likely insider trading—and thus illegal (Kiernan and Seaver 2017). Genetic models allow computers to learn to walk on two legs (Geijtenbeek, Panne, and Stappen 2013). So-called adversarial networks and various other types of neural networks can depict what a particular winter scene would look like in the summer, what a photograph would look like if Monet had painted it, and more (see, for example, Karras et al. [2017]; Metz [2017]; Zsolnai-Fehér [2017]; and Figure 2.3 on next page).

Photograph
Artistic Style

Artistic Style

D-Tech Composite
(Ostagram.me)

Figure 2.3: An example of computerized creativity. The computer was presented the photograph and artistic style, and developed a composite of the two.

D-Tech recently even developed intuition. We often think of computers calculating, "Option A will pay you $5, while Option B will pay you $3." However, we humans don't calculate everything in life. Ming, for example, recently graduated from college. Should she take the job with the startup paying a bit less or the job with the established Fortune 500 firm? Ming knows the starting salary and the job description but not how the job choice will impact her entire future career. Ming uses her intuition (and a lot of advice) to decide.

Google's Deep Mind developed software recently that mirrors some of Ming's thought-process. This software company developed Go Zero, a computer that applied intuition to dominate all challengers at the board game Go (Silver et al. 2017).

In Go, players alternate placing pieces on a 20 by 20 board of squares as shown in Figure 2.4.[9] If one player encircles the other's pieces, the encircled pieces are removed from the board.

The game sounds straightforward, but it's not: there are more possible moves in a game of Go than there are atoms in the universe![10] No computer could possibly calculate which move would be best. The *only* way to win is using intuition.

Figure 2.4: The board game Go, in which computers have recently dominated human competitors. Both computers and humans can now master the intuition required for competitive success at the highest levels.

Humans play Go using rules of thumb. For example, "If your opponent has many pieces in one quadrant of the board, then put your pieces in another quadrant." Go Zero developed innovative rules of thumb, providing insights from which even the best human players learned. Go Zero is far better than any human at Go: it won 100 games to zero against a software program that beat the human world champion, Ke Jie, 3 games to 1.

Yet D-Tech remains limited in its ability to analyze our environment. For example, D-Tech finds it more challenging to recognize the options to consider. Ming may decide to tour Europe for a year rather than taking either job. Today, D-Tech is unlikely to identify such free-flowing options.[11]

Nonetheless, D-Tech's ability to analyze our options is rapidly expanding. In terms of brute force computation and search capacity, D-Tech is already far beyond any human's capabilities. And, in terms of many other tasks, D-Tech is, step-by-step, increasing its abilities relative to humans.

Once again, D-Tech is not simply duplicating our human approach. D-Tech thinks (analyzes the world) differently from the way we do. The mistakes D-Tech makes illustrate this difference in thinking. For example, researchers changed a single pixel (dot)[12] in a picture of actor Sylvester Stallone—and caused a computer to think the picture was of actor Keanu Reeves (Mikhailov & Trusov 2017), as in Figure 2.5.

The researchers also changed a single pixel that caused the computer to think a picture of a dog was actually a stealth bomber, and that a different dog was a taxi. In fact, in 74% of pictures (Athalye 2010), there is a single, key pixel that can radically change what D-Tech thinks it sees. We humans don't change our perspectives that way.

D-Tech is also always rational. If you want temper tantrums, drug addiction, and sleeping with your boss's spouse, we need to get humans involved—or we need to program Digital Technology to do these things.

It's like we're in a cross-cultural marriage with D-Tech. We share enough commonality to work together, but our assumptions about the world are *very* different. We speak in words while D-Tech speaks in mathematical equations and data. Somehow, we need to find a way to work together, however, since divorce is not an option.

Sylvester Stallone [13]   Keanu Reeves [14]

Figure 2.5: Two very different-looking actors. When a single pixel was changed on Stallone, however, a computer now thought the picture was of Keanu Reeves.

Like a spouse undergoing counseling, D-Tech is getting better at sharing its thinking with us.[15] For example, have you ever heard a beer critic describe why she thinks a particular beer is good or bad? It sounds something like:

> *A nice dark red color with a nice head that left a lot of lace on the glass. Aroma is of raspberries and chocolate. Not much depth to speak of despite consisting of raspberries. The bourbon is pretty subtle as well (Lipton and Elkan 2016).*

That particular description was created by D-Tech explaining its thoughts about a particular beverage. However, D-Tech can apply millions of equations and billions of points of data to analyze particular situations. It is inherently challenging to summarize such associated "thought" in a way that will answer all of our human questions.

So what happens when we combine these decidedly power-ful—yet also non-human—approaches to gathering informa-tion, understanding how our world operates, and analyzing our options?

PUTTING IT ALL TOGETHER: ARTIFICIAL INTELLIGENCE

Increasingly, D-Tech is able to put all of the thinking pieces together: Gathering Information, Understanding Context, and Analyzing Options. Over the past few years in particular,[16] D-Tech has taken steps toward being able to use thinking to achieve more and more of the outcomes it desires on its own. We call a D-Tech tool capable of such success Artificial Intelligence (AI).[17] AI is making decisions about how to operate complex machinery, how to recommend television shows we may like, and how to set prices at the highest level we would be willing to pay.

So, where does that leave D-Tech? We're already comfor-table with D-Tech's ability to think more capably than humans in many respects. It adds numbers and searches text millions of times faster than we can. It can identify someone who is lying better than we can (Crowell 2015). Technology can confirm pneumonia on an x-ray better than can physicians (Kubota 2017). In fact, Google's chief futurist, Ray Kurzweil (2017b), forecasts we will have computers more capable of thinking than we humans by 2029.

No. D-Tech is *not* smarter than we are today, in general. The technology tends to be highly skilled only for particularly narrow slivers of the real world, with many applications over-hyped. The x-ray viewing Digital Technology mentioned above, for example, is better than humans at interpreting frontal x-rays. Yet, inter-preting frontal x-rays is only a sliver of the task performed to diagnose pneumonia. A physician analyzes x-rays from multiple angles, considers the patient's history, and interacts with the patient (Rajpurkar et al. 2017).

ACCELERATING OTHER TECHNOLOGICAL DEVELOPMENT

The impact of D-Tech on technology has gone far beyond expansion of new Digital Technologies. This expanding impact is similar to the impact of Agricultural Technology thousands of years ago. The impact of Agricultural Technology extended far beyond the relationship between humans and our farms. Agricultural Technology spurred the development of settlements, which promoted the development of professions as diverse as shoemakers, doctors, and carpenters.

The impact of D-Tech on technology encompasses *all* technology. Earlier, I noted that, with the help of D-Tech, we publish 2.5 times as many journal articles *every year* as existed *in total* in 1950. We publish 40 times as many books (over one million [Piersanti, n.d.]) as we did each year in the 1960s (Greco, Milliot, and Wharton 2014, 35–36). This information is innovation's raw material—and it is feeding a ravenous machine.

Mr. Kurzweil's forecast that computers will be smarter than we are by 2029 may not come to pass. Either way, we are in the midst of a spectacular D-Tech Revolution. D-Tech provides us increasingly sophisticated access to more and more information. It can influence our understanding of context in many ways, and it can analyze our options according to cold, calculating criteria. That's powerful. And when packaged as Artificial Intelligence that's just staggering. Moreover, as we use D-Tech to extend the capability of human thought alone, we are accelerating development of all technology.

Understanding D-Tech's *direct impacts* is an important exercise. Constructively, we can automate factories, install cameras to monitor our children's safety, and search for Government's latest regulations.

Destructively, we could automate warfare ("Slaughterbots" 2017). AI could eliminate 1/3 of current American jobs by 2030 (Franck 2017). Black hat programmers could steal money from your bank. One bank robber was recently arrested for stealing over one billion dollars (D. Meyer 2018). When I was growing up, the imaginary evil villains in James Bond films couldn't even steal that much! Other books and articles discuss these direct impacts well.

Understanding D-Tech's *indirect impacts* or *side-effects* is just as critical as understanding D-Tech's direct impacts. We are not operating today's computers in the world that existed in 1970. We are operating today's computers in a completely different environment. Let's look at how applying D-Tech has transformed that environment.

HOW IS DIGITAL TECHNOLOGY CHANGING OUR ENVIRONMENT THROUGH INDIRECT IMPACTS?

As we apply D-Tech, we are making our world more complex. We are frequently overwhelmed by detail, misled about how our world is operating, and make decisions with outcomes far different than intended. We discuss each of these impacts, in turn.

WHAT IS GATHERING INFORMATION LIKE TODAY?

Nikita's meeting with Dr. Suarez illustrates today's information gathering:

Nikita sits in a cold, sterile exam room. Dr. Suarez gave a short smile when Nikita entered. Now, Dr. Suarez's attention is focused on the computer screen. My grandparents would have looked at a picture of this scene and asked if Dr. Suarez was treating the computer rather than Nikita. Today, such a scene is the norm in a doctor's office.

When Dr. Suarez completes Nikita's appointment, the computer time continues. Dr. Suarez will spend twice as much time completing digital forms as she did with the patient, Nikita.

Where human connection once dominated, the focus on gathering digital information now takes the lead. We regularly face three types of challenges gathering the information we will evaluate:

- We are overwhelmed by the volume, detail, and level of technical content.
- We find ourselves awash in errors.
- We are spread too thinly by communications that promise relationship building.

GATHERING INFORMATION CHALLENGE 1: OVERWHELMING CONFUSION

We are enamored with D-Tech's ability to store and process information and connect people. As a result, we are creating massive pools, lakes, even oceans of information and options. We are reliant on D-Tech to help us search through the masses of options and detail. Especially where we are not specialists, however, our information search is likely to lead to confusing masses of data.

Figure 2.6: Human connection is challenging
to deliver in a modern doctor's office.

Our patient from the last scenario, Nikita,
experienced such confusion after her visit with
Dr. Suarez. To continue the illustration:

> *Unfortunately, Nikita is not well. She thinks about the final
> words from Dr. Suarez: "Nikita, you could die." That's all
> Nikita remembers from her doctor's visit. That and, "You need
> surgery. We could do angioplasty or a bypass."*
>
> *Nikita sits at her desk. Google gave her 14.6 million articles
> about her surgery options. Fourteen. Point. Six. Million.
> Articles! Nikita reads a handful. These articles don't simply
> compare two options—they are overwhelming. There are case
> studies, intervention studies, and malpractice studies. Nikita
> learns that she has more than two options—each of the surgical
> options Dr. Suarez mentioned has dozens of variants, each with
> pros, cons, and supporters. Nikita looks into her coffee cup and
> closes her eyes to block out the confusion. She has no idea what
> to do. She would have been better off with far less—and far
> simpler—information.*

Like Nikita, we frequently freeze in the face of overwhelming
confusion. It is as if the tsunami of details steals our sanity. We
freeze. We reach "cognitive and information overload." "The
brain region responsible for smart decision-making has essentially left the premises (Vaughan 2014)."[18]

Where do we find such a confusion of information?
Everywhere! During our everyday lives we run into one information overload after another. Over the course of a complete
lifetime in the 17th Century, the average person was likely to
come across less information than is contained in a weekday edition of the *New York Times* (Wurman 1989). During *every minute
of the day* (Kimmorley 2015):

- Email users send 204,000,000 messages. The average individual office worker receives approximately 140 emails every day (Radicati 2014).
- Google receives over 4,000,000 search queries.
- Facebook users share 2,460,000 pieces of content.
- YouTube users upload 72 hours of new video. To watch all of the video uploaded in a single minute would take you a full three days.

Where do we torment ourselves with overwhelming options, most of which superficially appear more differentiated than they actually are? Everywhere! Amazon offers more than 300 options for toothpaste: mint, peppermint, spearmint, cinnamon, Total White, Pro Health, 3D White, Complete, Fresh Breath, Deep Clean. And that's just the beginning.

Between 1975 and 2008, the number of products in the average supermarket mushroomed from 8,000 to almost 47,000 (Consumer Reports 2014). The number of truly unique foods has expanded—but far less than those numbers might suggest.

Today, experts say we suffer from data smog, or "data asphyxiation," "information fatigue syndrome," "cognitive overload," or even "time famine" (Levitin 2015). And such data smog is only getting worse. In 2010 and 2011, the world created nine times more data than had been created throughout all of human history to that point (IBM 2011). And that was just the beginning.

To find the information and options we need in this type of data smog, we *must* use D-Tech to gather, and filter, our information. However, in areas where we're not specialists, D-Tech's filters frequently don't help.

As a parent, I want a video for my kids that will both entertain and educate. Yet, I'm not an expert on education—I don't know how to precisely describe what I want. Instead, I type "Excavator video for kids" into YouTube and thousands of (generally medi-ocre) options appear.

Figure 2.7: The information and choices
on offer are overwhelming.

I can certainly search for "what makes a video educational for children?" However, the 455 million results simply overwhelm me. Like Nikita, I'll read a handful of those entries, but specialists are confusing! For example:

> Koumi (2014) presents a division of the specific pedagogic roles of video for 'techniques and teaching functions for which video is outstandingly capable' which is due to video's 'rich presentational attributes [which] can result in learning facilitation'. The functions are allocated into four domains: Cognitive, Experiential, Affective and Skills and a link is made to a revised version of Bloom's Cognitive Learning Taxonomy. (Woolfitt 2015).

What did that say? "When a teacher uses a good video, kids learn more."

In short, with D-Tech we have come to live in a virtual world in which data is king, reigning over our tasks for Gathering Information. We strive to create more data yet can also feel overwhelmed by it. We use that information to create and track ever more options—that do not necessarily add more value to our living experience. We use D-Tech to filter the masses of information and options. Yet, especially in areas where we are not experts, we become lost.

Such a characterization doesn't discount the power of D-Tech's information filters. Where we are experts, or need expert information, D-Tech consistently gives us what we want. What is the impact of fatty acid C-16 on a turkey's liver? *Got it!* Oooh. Not good. What was the name of Jefferson Airplane's lead singer? Google just told me it was Grace Slick.

Nonetheless, when we need helpful information about how to actually live our lives—too frequently we are lost. Is a chicken pot pie or a roast beef sandwich more healthful for me? D-Tech doesn't help me come to an answer. In fact, D-Tech's overwhelming data on nutrition simply confuses me.

In addition, even when we find the information for which we're searching, D-Tech sometimes still gives us erroneous data. What's the problem?

GATHERING INFORMATION CHALLENGE 2: ERRORS AND MORE ERRORS

Too frequently, D-Tech gathers error-filled information for us. We humans create much of the mess by accident.

> *Dr. Suarez, the physician treating Nikita, tries to minimize her mistakes, but they appear nonetheless. The average medical record today contains eight errors (Bowman 2013). Impacts from soiled data, unfortunately, arise.*

Patients have missed years of potentially life-saving cancer treatments due to data mistakes: a software bug pulled up old test results rather than new ones (Singer 2010). An infant suffocated from a massive overdose when a handwritten order was mis-transcribed (Versel 2011).

We have always been error-prone. Yet, by stressing our ability to focus and respond rapidly, we are increasing the data errors D-Tech pulls up for us—of our own creation. Then, D-Tech, itself, compounds the mistakes by hiding these errors in data-lakes—or what we might call, "Data Cesspools."

I'm human. I made eight typing mistakes in the past two minutes. The backspace key is my best friend (it's the third most-used key on the computer [Yarow 2013]). And, I left two grammar mistakes in the writing that I hope the editor catches later. Maybe I'm worse than most, but I don't think so.

When we focus on complex tasks such as writing, we focus our brains on high-level activities like conveying meaning. We focus less on low-level activities like grammar and spelling. As a result, we don't catch every detail of lower-level thinking. The higher the level of thought, the more likely we'll make mistakes in lower-level actions. For example, we misspell "if"; we mix up our children's names; we add an extra "not" in a sentence; or we paste the paragraph about ocean vessels in the page about Norwegian skiing. In today's complex world, we have extremely little brain power left for lower-level activities.[19] Such errors may not cause severe problems when I'm writing a book; however, when Dr. Suarez makes such errors, she's threatening lives.

We also tend toward careless communication with D-Tech. In principle, we know we can later correct any mistakes we make with D-Tech. In practice, however, such mistakes remain etched in our virtual record, left to sow the seeds of confusion later. Once such errors exist, D-Tech tends to disguise them. Businesses have created so-called "Data Lakes" that disguise masses of errors in even bigger masses of information.

Hidden in the Data Lakes is a mess of "poor data hygiene" (Dignan 2017). What's poor data hygiene? We could ask Frans, a data guru I worked with. Frans would pull up a weekly report we'd run to track next week's deliveries. Most weeks, the report would project a total of about 3,000 tons of deliveries to customers. Every once-in-a-while, the report would be much smaller—sometimes only a few hundred tons.

Frans would explain, "Our factory database has a mistake in it. Someone long ago recorded that the last day of the month will always end a week: this report only captures production for Sunday the 30th and Monday the 31st. The plant will operate the other 5 days of the week—but poor data hygiene allows this little mistake to remain years later." As the saying goes, "Poop in. Poop out." And far too many data lakes are excrement-filled.

Figure 2.8: Many data lakes deserve the nickname data sewers.

Recently, businesses and technologists are making greater efforts to improve data hygiene. There are new methods to check self-driving car software for bugs, for example (see Kurzweil [2017a]). Software platform vendors (such as Microsoft and IBM Watson) pitch their own sets of tools to clean data that are particularly effective for clients who only use that vendor's tools. While data hygiene efforts have improved somewhat, realistically the hard work of maintaining data integrity (i.e., "good data hygiene") depends, at its core, on human practices. Unfortunately, "Humans aren't up for the job and never have been . . . Humans aren't meticulous enough (Dignan 2017)."

Bottom line: An ocean of complex, nuanced information is an invitation to error creation, and to machines hiding the problems. D-Tech brings its own ability to reduce errors, but our human tendencies are battling that capability. There is truth to the saying attributed to Mitch Ratcliffe (1992), "A computer lets you make more mistakes faster than any other invention in human history, with the possible exceptions of handguns and tequila."

D-Tech gathering can also provide us the wrong information for human relationship development. Rather than helping deepen meaningful human connections, D-Tech tends to destructively deliver a too thinly spread set of information. Such information is driving us to hollow out our relationships. Let's discuss this in the following section.

GATHERING INFORMATION CHALLENGE 3: RELATIONSHIP-DEVELOPMENT DATA

What is emblematic of our relationships today? Let's go back to Nikita again—she's crying at her desk *alone*:

Nikita is experiencing one of the most painful periods in her life. She might die. And, yet, she's alone. Why? Nikita would feel uncomfortable discussing her life-threatening illness with anyone she knows. She used to have a deep bond with her younger brother, but they've grown apart over the years. Nikita's dad has passed away and her mom is too frail to provide her strong, emotional support. Nikita's single, and she doesn't have deep friendships.

Nikita may be despairingly alone, but she's not unique. With a click of a button, we invite people we just met to be "friends" or "connections" that will be digitally stored forever. Yet, these connections are not friends. These connections are generally thin, digitally facilitated records of our past experience. And, our focus on such thin connections is hollowing-out our relationships.

I don't truly have 500 friends through Facebook and 1,000 relationships through LinkedIn. Instead, I have that many electronically supported connections. The human brain, however, is only large enough to maintain 150 relationships (Dunbar 2011). Rather than 150 meaningful relationships, we have hundreds or even thousands of hollowed-out, digitally supported connections.

As we'll discuss in Chapter 3, the direct implication is a loss of connection to other humans. We're feeling more lonely and alone.

There's more. With D-Tech, people physically close to us have lost their prominence. Out of sheer necessity, most of our close relationships used to be with those in our neighborhoods. As we developed these relationships, we became better able to work together to solve neighborhood issues. We understood our neighbors' families and businesses. We understood the implications of their desire to build a store next to our home, or we could understand the impact of accompanying them on a particular doctor visit. We collaborated effectively with each other, in everything from building permits to love, life, and loss.

Through D-Tech, vast numbers of people far from us have the same access as those close by. There are certainly electrifying benefits from being able to develop relationship with people physically far away. Only through D-Tech could I romance my future wife from half a world away. We initially met on a Kuala Lumpur airport train in the physical world. We still met for breathtaking dates to the Louvre and the Roman Coliseum, "halfway between her house and mine." However, we maintained our relationship through D-Tech's free texting, calling, and video-chats. Even 50 years ago, our romance would have been "virtually" impossible; our spouses would have been far more likely to come "from the next village over."

I have a stunning wife from the other side of the world, but I barely know my neighbors. As we'll discuss in Chapters 3 and 5, the lack of neighborhood connections is impairing development of our neighborhood businesses, construction, and even research. Such impairment is more impactful than it may seem: much of our world remains associated with our homes and neighbors. As we'll discuss in Chapter 5, the U.S. economy may be 40% smaller today as a result of these neighborly connections being weakened.

As we gather information in our D-Tech world, we are overwhelmed by a confusing sea of information. We are dependent on D-Tech to filter that information, yet these filters are relatively ineffective at finding us critical information, especially in areas where we are not specialists. Much of the information D-Tech does retrieve for us is awash in errors—and the information it retrieves about relationship-building is spread too thinly for us to develop truly deep friendships.

Once we have information gathered, we evaluate it to understand how our world operates. In critical respects, this step is also more challenging in our D-Tech world.

WHAT IS UNDERSTANDING HOW OUR WORLD OPERATES LIKE TODAY?

There is a control room with 100 people, each with little dials . . . This actually exists right now . . . A handful of people at a handful of companies will steer what a billion people are thinking today (Harris 2017)."

The people in this room will control thinking by managing how our social media work. The people in the control room, and their software, feed off data—and those people are getting fat. Mireille, a user of Tinder (a dating application), requests the data stored about her—and finds 800 pages of information about herself. Everything from her locations, sexual preferences, jobs, pictures, and musical tastes is recorded. So is the joke Mireille copy-pasted to potential dates number 567, 568, and 569. Her lazy, almost automated approach to flirting is captured in that repeated joke. Mireille cringes when she sees the recording of that flippant action (Duportail 2017).

Anyone who wants to manipulate Mireille's emotional response to the world has an arsenal of weapons in such records. The people in the corporate control room can "personalize the experience,"[20] or, as Tristan Harris (2017) describes, "give the little dials power."

We humans may be the most intelligent animals: we create sophisticated stories about how our world works when given even the most basic information about our surroundings. Yet, this ability to construct sophisticated understanding from even limited data has a downside: we can be manipulated. As

Fauconnier and Turner (2002, 17–38) explain, the way in which we are presented information actually shapes how we fill in the blanks and make sense of that information; context is everything.

Our ability to be manipulated can make us look stupid—even less intelligent than a rat. Psychologists pitted Yale undergraduates ("Yalies") against rats—and the rats won (Liberman 2005).

The psychologists randomly chose to place a pellet to the left or right, with right three times as likely to be chosen as left. Before seeing where the pellet was, subjects (human or rat) chose a direction. A subject that chose the direction matching the pellet was awarded a point on the scientist's scorecard.

The rats beat the Yalies. Rats rapidly learned to scurry right—always. That strategy is by far the best strategy for the game. The Yalies, in contrast, chose left sometimes, well into the game.

Understanding why the Yalies lost provides insights into how we think. The rats were playing the psychologist's game: find the pellet.

The humans, on the other hand, were likely filling in the blanks to play a more complex game. When scavenging in a tribe, the winning strategy includes sometimes searching for food in a smaller (less likely to win) field. Sometimes you, as an individual, win—because no one else goes to that smaller field.[21] It's likely that the Yalies were plugging the simple data from the psychologists' game into such a complex, nomadic survival game (Liberman 2005.) The Yalies were essentially manipulated into losing to rats by a game too simple for the Yalies' intellect.

It's not just Yalies who can be forced into misunderstanding how our world operates. We all can. There are at least 188 ways we can be manipulated into misunderstanding how our world is operating (Manoogian and Benson 2016).

Experts call such approaches *cognitive biases*: tendencies for humans to fill in the blanks around data in particular ways. Want to sell your book at a higher price? Get the customer's brain anchored on a big number like the distance to the sun. Then show the customer that your price is lower than the anchor; the customer's brain is wired to think of this price as inexpensive.

There's no precise path to influence humans—we're never exactly sure how big of an impact we'll have on another person's thinking. Yet, with our large toolbox and with enough attempts, we're likely to have an impact.

With D-Tech, we've multiplied the potential to impact "how we fill in the blanks." We face new horizons in our drive to "nudge" or influence others (Veltri 2015). We have:

- Strengthened our ability to manipulate our perception of the world around us.
- Weakened our ability to control how D-Tech manipulates our perceptions.
- Extended the reach of any manipulation that occurs.

UNDERSTANDING OUR WORLD CHALLENGE 1: ENHANCED ABILITY TO MANIPULATE

In that control room, 100 people with 100 dials are steering how a billion people will think today. (Shudder)

D-Tech is a powerful manipulation tool. Want me to pay attention to my phone no matter what else is happening around me? Make the phone *ping* when a message arrives. Want me to pay attention to my robot? Have the robot pretend to feel sad. Want me to drink a coffee? Show me a coupon for that coffee the second I walk in front of the coffee shop.

What's the difference between such D-Tech manipulation and traditional advertising (which also aims at controlling our understanding of the world)?

"Start 'er up and mow 'em down!" That's a human starting the tractor that will devastate an entire forest. We have always been able to pull a single tree down by hand, but the impact on our environment was modest. With tractors, we can demolish entire continents of forests.

D-Tech-driven influence has similarly greater power than does traditional advertising. Through D-Tech, we have access to:

- Every gram of reference material available about how to influence humans.
- Every ounce of data and history that defines us as individuals.
- The ability to communicate to the target anywhere and anytime the target appears susceptible.

D-Tech-driven influence isn't just advertising—it's automated manipulation.

As we will discuss further over the next chapters, the results are already concerning. In some cases, options are simply taken from us, such as from the Rohingya ethnic minority in Myanmar. In many areas of Myanmar, Facebook is the only news source, and it "is currently being used to spread dehumanizing misinformation about the Rohingya." As a result, Facebook is blamed "for more than half a million members of this ethnicity being driven from their homes due to systematic murder, rape, and burning (Thomas 2017)." This minority group does not have the option to protect themselves, their family, and friends.

In other cases, we are being actively manipulated to believe an exaggerated importance of D-Tech's desired actions. The recent robotic toy, Cozmo, for example, is described as, "a robot capable of playing on a young kid's emotions." "If [someone] neglects the robot, [they] feel the pain of [the toy expressing fake sorrow]." Children exposed to Cozmo presumably heed this need to interact with the robot. The impact? "Such robotic training may leave children without the equipment for empathic connection" with other humans (Madrigal 2017).

Many similar situations exist for American adults. Russia, for example, has acted to inflame our political choices based on Russia's own desires. Anyone influenced by the following digital advertisement, Figure 2.9, was manipulated by Russian politics, not American (Kang, Fandos, and Isaac 2017).

Figure 2.9: A political advertisement for the U.S. posted on behalf of the Russian government in 2016.

D-Tech-driven influence is not all-powerful. Technology's attempts at influencing us humans are frequently unsuccessful. Attempts to influence us to exercise and lose weight through the use of wearable technology have largely failed, for example (Jakicic et al. 2016). Yet, at least in theory, D-Tech provides us greater potential to influence our understanding of the world around us than any tool available to the fascists and ethnic cleansers of World War II.

The power in that control room is formidable. In the examples above, D-Tech's goals are straightforward: convince viewers the Rohingya are horrible people, persuade children that a robot is sad unless the robot receives attention, and win over voters towards a particular candidate. Each of those goals is independent of the user's current situation.

On the other hand, D-Tech use is frequently sensitive to the user's current environment. For example, the movie D-Tech suggests we watch depends on the movies available. In such situations, our ability to control user manipulation is weakened, as we discuss now.

UNDERSTANDING OUR WORLD CHALLENGE 2: WEAKENED ABILITY TO CONTROL MANIPULATION

If you only looked at Tinder data for Mireille's three most romantic dates ever, you'd walk away breathless. Perhaps you'd think Mireille had superhero romantic powers. On the other hand, if you looked up Mireille's three least romantic dates, you'd find 14 impatient sighs and 38 yawns.

The data we receive tell a story about how our world operates. More and more frequently, we are relying on D-Tech to tell us that story. We rely on our programmers to fully understand how the software will be used along with creative programming to ensure a constructive story is told.

Consider Meetup, an application designed to help people find group social activities. Meetup's software shows users different meetings in which they are likely interested. While men do tend to prefer tech meetups more than women do, a so-called neutral software could easily run all women out of technology meetings:

> *Such a neutral software would show Dave more tech meetup options than it would Rosa. As a result, Rosa would attend even fewer tech meetups—she wouldn't even realize such meetups exist. Rosa's lack of attendance would reinforce the software's expectation that women don't like technology. It would show Rosa and other women even fewer tech options as a result, in an out-of-control spiral. Only by actively countering this women-in-technology bias does the company Meetup stop such an out-of-control bias spiral from occurring (Estola 2016).*

In other words, the only way to ensure D-Tech appropriately informs us about our world is to creatively program it to do so. There is no such thing as a neutral technology platform—because a neutral software would expand every difference or bias that already existed. As iRobot director of Data Science Angela Basset (2017) explains, "It's not that data can be biased . . . Data IS biased." The data we see reflects the context in which it was created.

With D-Tech involved, a human no longer has direct control of the story being told. Instead, we must rely on indirect control: a human programmer must understand how the software will be used and creatively design the approach so the technology tells the desired story.

Where programmers fail in their delicate tasks, D-Tech can warp the message. Google, for example, recently offered access to a racist, anti-Semitic "sentiment analyzer." Judging whether a million tweets about a product are, on average, positive or negative, for example, is the tool's strength. Type in, "That's a great soap!" and the analyzer tells you that phrase communicates positive sentiments.

However, the Internet trained Google's sentiment analyzer. Anything the Internet judges to be negative would be identified as negative. Black, female, Jewish, and homosexual were all judged as negative sentiments. Christian and even "white power" were positive (Thompson 2017).

In a 1960s movie, an actor would be slapped for saying any of the overtly racist and otherwise destructive statements the Google analyzer judged as positive. What happened to the software that made equivalent statements? Did the software programmer get slapped?[22] Probably not. As we lose the ability to directly slap (provide feedback) the human involved—we weaken our ability to control warped messaging.

UNDERSTANDING OUR WORLD CHALLENGE 3: EXTENDED REACH OF ANY MANIPULATION

Mireille learned the hard way. She'd provided an ex-boyfriend a slightly risqué picture of herself. That picture is now posted on 143 different Internet sites. When Mireille interviews for a job, she knows the human resources representative has almost certainly seen that picture of her in her underwear. It's not the end of the world for Mireille—it's just an added reason to be nervous in a job interview.

Once created, the data has a life of its own. Automated bots can post information on the Internet hundreds or even thousands of times. And we humans tend to multiply such destructive information. Ever heard:

"You only use 10 percent of your brain." "Eating carrots improves your eyesight." "Vitamin C cures the common cold." "We have high levels of violence against law enforcement in the U.S."

These statements are all false. They sound true because we've heard them over and over. This tendency is called the "illusory truth effect." We tend to believe and pass on false information we see frequently enough (Dreyfuss 2017).

With D-Tech, this tendency toward misperception of our surroundings becomes super-powered. Recently, a study (Lazer et al. 2018) found "a false story reaches 1,500 people six times quicker, on average, than a true story does. And while false stories outperform the truth in every subject—including business, terrorism and war, science and technology, and entertainment—fake news about politics regularly does best."

Consider two examples from the 2016 election. "In August 2015, a rumor circulated on social media that Donald Trump had let a sick child use his plane to get urgent medical care." The story was confirmed as true. Nonetheless, "only about 1,300 people shared or retweeted that story (R. Meyer 2018)."

On the other hand, "In February 2016, a rumor developed that Trump's elderly cousin had recently died and that he had opposed the magnate's presidential bid in his obituary. 'As a proud bearer of the Trump name, I implore you all, please don't let that walking mucus bag become president,' the obituary reportedly said." This tale was found to be false. However, "roughly 38,000 Twitter users shared that story (R. Meyer 2018)"—more than 15 times as many people as shared the true story.

Perhaps we simply find false information more appealing to share with others than we do true information. Whatever the case, our tendency to propagate false information is becoming increasingly hazardous as D-Tech super-powers our ability to replicate false information.

The information we use to interpret how our D-Tech world works frequently misinforms us. We've super-powered the ability to influence our understanding of the world in a way that favors specific, concrete outcomes (such as, pay attention to your phone.) At the same time, we've weakened our ability to ensure D-Tech appropriately informs us about our current, changing environment. And, we've extended the reach of any inappropriate perception or manipulation.

If we've reasonably successfully understood how our world currently operates, we have one more hurdle. We must decide what to do—which involves anticipating how the world around us will change depending on what we do. In critical respects, this step, analyzing our options, is also more challenging in our D-Tech world.

WHAT DOES ANALYZING OUR OPTIONS LOOK LIKE TODAY?

New janitorial standards almost killed 78-year-old patient Betty. How did that happen? Unintended consequences:

> *Well-intentioned new regulations were introduced to reduce false-positive tests for Gramm Negative Rods (GNR): "a class [of bacteria] that includes E. coli, Salmonella, Shigella, Klebsiella, Pseudomonas, among others, and that has within its lineup a murderer's row of multidrug-resistant strains . . . GNR is as common in hospitals these days as red Jell-O (Leaf 2017)."*

> *In fact, it was a false alarm that brought Betty to the hospital in the first place. Betty's physician, Dr. Rosen, had called Betty saying, "Betty, you may have a life-threatening infection. Your blood work tested positive for GNR. Return to the emergency room immediately!"*

> *We certainly need fewer false alarms like that!*

> *The regulation seemed like such common sense: require just a bit more cleaning of sample testing rooms to reduce the GNR growing there. Fewer false alarms. Better healthcare.*
>
> *However, ripples are set off with such regulations. Supervisor Harold needed to redeploy his hospital janitorial staff to satisfy the new regulation. As a result, the janitor Luisa wasn't where nurse Nancy expected her to be. Nancy spent an extra 20 minutes searching for Luisa, which caused Nancy to be rushed on her hospital rounds—and make a life-threatening mistake handling an 78-year-old patient.*

D-Tech has brought us unrivaled capacity to analyze decisions—and we do lots of such analysis! Our insights are deeper and more poignant than ever before. D-Tech exposed the commonality of false GNR test results, after all. Yet, as we learn more, we're creating complex, interrelated systems—with more and more different types of specialists. A nurse in a modern hospital, such as Nurse Nancy, regularly relates with so many different types of physicians, technicians, and other white collar, blue collar, and pink collar specialists.[23] That's staggering.

For positive outcomes, we need all of these different specialists to align their decisions. A change in the behavior of any one of those types of people can ripple through the entire system. A regulation in how the janitorial staff spends their time impacts the interaction between nurse and patient? In today's D-Tech world? Certainly! And that's just the start: unanticipated consequences are the norm, not the exception.

As we adjust the decisions and actions of different actors in such systems, we are creating what are best-called *legacy systems*: brittle, ill-fitting systems that do a better job of representing a system's history than its present.

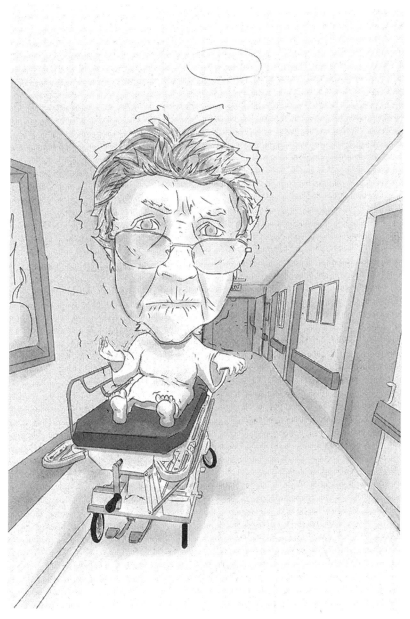

Figure 2.10: 78-year-old Betty is having
a challenging day in the E.R.

Patel (2017) described information technology (IT) legacy systems for the Financial Times, but his description applies equally to legacy systems across society:

> *Legacy systems are a reflection of [the] past and present; they mirror both the complexity of the world they were developed for and that they currently operate in. If you peel away a system's layers you see [increasingly ancient rules that should be in the history books, yet still determine how the system operates.]*

The history of such systems is understandable. Most systems begin small:

- In the 1880s, Thomas Edison created the first electrical grid to power only one square mile of New York City (Campbell-Dollaghan 2015).
- The Apple computer company began with two guys working in a lab and a couple of others being unproductive in Steve Job's garage (Wozniak 2014).

Each small system has its own unique approaches and needs, perhaps a unique currency, tax, or type of electrical wire. Over time, these small systems are combined and expanded, cobbled together and patched. The company grows, divisions merge, and an entire city needs electricity. Requirements change as the electric system needs to power subway cars as well as light bulbs.

Once systems grow to touch many different activities, they become incredibly complex (Pega, n.d.). Many Information Technology (IT) systems are legacy systems. Of corporate IT budgets, 72% are targeted at cobbling, patching, and otherwise simply keeping the lights on (Zetlin 2013). For 75% of government IT projects, *all* of the funding is dedicated to such activities (GAO 2016).

Like Supervisor Harold and Nurse Nancy, we begin to make unaligned decisions in such legacy systems and our collaboration begins to fail. For example, Delta Airlines cancelled 2,300 flights and posted a profit warning after a backup reservation system didn't kick in when it should have. And Deutsche Bank's CEO John Cryan described his bank's IT system as a "Horlicks": a total mess. Because of this mess, this Fortune 500 financial institution is unable to effectively manage financial risk and execute its financial trading (Patel 2017).

Through D-Tech, we're creating far more complex and challenging to manage legacy systems:

- Each system has more parts and specialists requiring decision-alignment.
- Our systems are becoming increasingly brittle—with un-anticipated impacts becoming increasingly destructive.
- Our systems are operating increasingly rapidly—reducing the opportunities to watch what's occurring before action is necessary.

ANALYZING OPTIONS CHALLENGE 1: MORE ELEMENTS REQUIRE ALIGNING

Why does Nurse Nancy need to interact with 130 different types of people? Even for a single human organ, the human heart, Nurse Nancy must interact with two different types of specialists: Dr. Kushwaha, the echo-cardiologist who listens to the heart, and Dr. Bashir, the electro-physiologist who measures the heart's electrical currents (Price 2015). If Nurse Nancy's group covered the entire body with every specialty, the physicians, alone, would comprise 131 types of people. (U.S. doctors can become board certified in 46 specialties and 131 sub-specialties [American Board of Medical Specialties, n.d.].)

Our ever-expanding knowledge demands such specialization. We have specialists in jet engine vibration, we conduct in-depth studies on the impact of animal feed on egg yolk color, and we spend months optimizing the size of the fat bubbles created by soup cubes. Across the sciences and arts, there are 633 majors in which a college student can obtain a degree (Chrisomalis, n.d.).

According to Glenna Crooks, author of *The NetworkSage* (2018), operating a small business today can require relating to 150 *types* of people,[24] while operating a modern, single-woman's household can require relating with 139 more. Each of these specialists speaks their own language and has their own way of looking at the world. And all of these people must effectively collaborate for the best outcomes.

D-Tech is similarly expanding the number of non-human elements in our systems. The windows we have in our home aren't single panes of glass. A cross-section of a modern window contains 67 labeled parts (Farooqi, n.d.). The average smartphone has over 300 parts (SourceMap 2015) and is covered by 250,000 patents (Masnick 2012).

Any time we redesign any of those parts or change the activities of any people, we need to change how a dozen others operate. Redesign 1 of the 67 parts in your window and 9 teams on 4 continents must immediately begin redesigning a dozen other parts.

ANALYZING OPTIONS CHALLENGE 2: INCREASINGLY BRITTLE RESPONSE TO CHANGE

Nurse Nancy has access to hundreds of pages of official processes and protocols. All of these systems are designed to mitigate problems.

Yet, our systems remain unstable. Nurse Nancy almost killed a woman by accident. Many nurses today spend the bulk of their days tired and angry. Why? Too frequently, the bulk of a nurse's day is spent fixing problems.

Today's challenges come from so many starting points and in so many forms that problems frequently don't fit within official processes. And, when problems do occur, the implications can too often become oversized.

As we apply D-Tech, we are making it more challenging to align. How do firms offer us 300 variants of toothpaste? For the most part, the difference between these offerings are miniscule—yet must be maintained.

Take Palmer and Greeley's recently introduced "All-Natural Toothpaste that Whitens Your Teeth with no Whitening Agents Added."[25] The paste is like any other toothpaste with one exception. Palmer and Greeley concentrated a special micro-molecule naturally occurring in 100 tons of fluoride into 5 tons of their new toothpaste.

Toothpaste manufacturers used to ignore those micro-molecules. However, when concentrated, the extra acidity will whiten your teeth. The molecules don't need to be labeled because they naturally occur in fluoride.

By introducing this one type of paste, Palmer and Greeley made a micro-molecule that used to be irrelevant now matter. Before this toothpaste type's introduction, any manufacturing change that reduced the amount of this micro-molecule would have been acceptable. Now, such a change is off-limits.

By the time we create 300 different toothpaste variants, almost any miniscule change to any manufacturing process matters in some way. We're hamstrung—unless we coordinate multiple, simultaneous changes, i.e. collaborate.

Every product is being similarly modified. So is every service. And even every *aspect* of society. The way we find new music to enjoy. The way we maintain voter registration lists. The way we reduce the number of blemishes on our tomatoes. And the number of Android phone and tablet models? There are at least 24,000 variants (Mirani 2015).

Alignment along digital specifications is even more challenging than are physical product or service specifications. Let's look at Airbus, the airplane manufacturer. Two Airbus engineers, one in France (Herve), and one in Germany (Heiko). Each had to decide whether or not to upgrade his engineering software. Technically, it didn't matter whether either one upgraded their software. We all sometimes upgrade, and sometimes we stick with what's worked in the past. However, it was important that the two engineers both make the same choice. They didn't.

Herve upgraded his version of the electronics design software to include 3-dimensions, while Heiko stayed at 2-dimensions (Greising 2006). As a result, the front half of the A-380 plane designed in France couldn't be fit together with the back half designed in Germany. Thousands of hours were spent trying to un-jumble 530 kilometers of electrical cables across the two airplane halves. One design team missing one dimension led to a waste of $6 billion (Reddy 2014). Oops.

The digital challenge lies in the fact that *everything* must be aligned digitally. In the physical world, I am me: two arms, two legs, and a head. An airplane always has three dimensions. Nature has pre-defined these objects for us. For digital specifications, we humans create every aspect of every definition. Different people are defining these objects at different times. And alignment is frequently missing.

What happens when such people and parts misalign? Failure. For example:

- An estimated 37% of business software is never used ("The Real Cost," n.d.). The software's generally fine, and the employees who are supposed to use the software are capable. However, no one effectively aligns the people's training with the introduction of the new software.
- As D-Tech helps us track complex, global supply chains, decisions about factory locations ripple across continents even years after decisions are made. A single flood in Thailand forced a slowdown in car factories across half the world (Fuller 2011).[26] All of the competing suppliers had located their factories physically close to each other, and the supply chain failure occurred, as a result, years later.

Moreover, as Digital Technology plays an increasing role in our world, the impact of our failures grows. Data breaches, for example, allow sensitive information about us to fall into the wrong hands. These breaches existed as early as the 1980s (Hayden 2013). Withal, their frequency is now exploding. In 2005, 157 data breaches were reported. By 2015, that had increased more than 5-fold, to 783 (Lord 2018). A single, recent breach laid bare information on 143 million different Americans—almost one in every two (Gressin 2017).

We can already see what the final state of this challenge looks like. For a self-directed, self-learning computer ("Artificial General Intelligence" 2018), even one, small, virtual-world error could have the ultimate consequence in our physical one—man against machine. Such an ultimate computer, if carelessly instructed to say, "Win more chess games" thinks that winning chess is its only goal. It realizes that money and other resources can help it enter chess tournaments and buy books—and thus the computer tries to dominate the world's economy. The computer realizes that by duplicating itself, it can play twice as many chess games. Thus, it decides to repeatedly duplicate its world-dominating self. And, the computer realizes the simplest way to win chess games is to remain turned on, and thus dedicates itself to protecting its off-switch from you and me (Omohundro 2008).

Computers developed by leading technology companies aren't coming after us tomorrow. Yet, we are living in ever-more brittle legacy systems. Coordination challenges are increasingly leading to unanticipated consequences—many of them problematic. And, the impact of our mistakes continues to grow, day-by-day, ever larger.

Our need to align is also growing as the world around us accelerates.

ANALYZING OPTIONS CHALLENGE 3: ACCELERATING SPEED

Nurse Nancy feels like I often do: constantly pressed for time. Our to-do lists seem to grow every day despite our efforts to get ahead. Technology is accelerating. Our ability to apply that technology is accelerating. Change is accelerating. And that change increases the challenge coordinating across our complex systems.

Perhaps complexity wouldn't matter if we had plenty of time to study exactly what our current world looked like before changing anything. If we had the time to discuss and coordinate, we could do so at leisure. However, we don't.

As a teenager, Pete amassed a sizable collection of vinyl records. Then he was forced to upgrade his vinyl records to cassettes. Then he was forced, yet again, to upgrade his collection from cassettes to CDs. After his upgrade to CDs, Pete threw up his arms and cried, "Can we all agree this will be the last form of music media we have to buy?"[27]

No, Pete, we can't agree to stop moving forward. In fact, we've already moved on to digital downloads. The world is changing. Fast!

Our collaboration today requires aligning broad numbers of people and parts that are each changing faster than ever before. Each decision-maker has access to D-Tech to analyze their individual decisions.

Yet, the key to success is aligning with all of the other decision-makers. Today, D-Tech's information helps us coordinate the most superficial of our decisions. D-Tech can, for example, ensure that there is always a nurse assigned to the emergency room. Today, however, D-Tech also lacks the ability to help us coordinate our more nuanced, strategic decisions; humans must align such decisions.

Perhaps digital decision-makers will replace human decision makers in the future. Yet, even then, aligning separate decision-makers will be critical. We earlier talked about the game of Go. Even Digital Technology must use intuition in playing Go because calculating every option is impossible.

Go is a complex game. But it is also just a 400-square board game. One human body, on the other hand, has 37 trillion cells, and we can think of each cell as a square in a hyper-complex board game. There are 7.6 billion humans on earth ("World Population" 2018) not to mention 10,000 trillion ants (Moore 2014), 230,000 species of ocean life (Jha 2010), and more stars than we can count in 50 years of summers.

Individual D-Tech machines will be unable to make all of the decisions about the world. D-Tech will always have such limitations despite the ongoing efforts to expand data storage and accelerate calculation. We will always need to coordinate across multiple decision-makers if we want the most constructive outcomes.

SUMMARY: THE ORIGINS OF DIGITAL TECHNOLOGY'S IMPACTS

In many ways, we have come full circle from the Agricultural Revolution of several thousand years ago that we discussed in Chapter 1. Agricultural Technology only directly impacts our ability to feed ourselves through the soil. Yet, applying that technology changed how society worked. The minority of Agricultural Revolution societies that succeeded in turning farming technology's potential into better societal outcomes evolved. These successful societies replaced their old structure of nomadic generalists with markets featuring specialized settlers.

Only those societies that evolved obtained the benefits that agricultural technology promised: a stable food supply which freed people to develop stronger social connections, enhanced self-esteem, and the prospects for self-actualization. Citizens benefited from everything from wondrous crop yields to philosophy and poetry. Those societies that did not evolve stagnated.

There are many parallels with our Digital Technology experience of today. The direct benefit of Digital Technology is applying a decidedly non-human approach to supporting our specialists by delivering these benefits:

- Unparalleled access to information and people.
- Exceptional ability to deliver insights into how our world is operating at a detailed level.
- Unrivaled capacity to analyze how individual decisions will impact our world.

We are combining those capabilities to create Artificial Intelligence, capable of achieving goals on its own and to accelerate development of all technology.

Application of Digital Technology, like Agricultural Technology before it, is also creating *side effects*: our world has become more complex. We are now skilled at supporting specialists—but find new challenges in bringing the benefits of those specialists to the rest of society:

- We are dependent on D-Tech to find the data we need in the sea of information we've created. Yet, in areas where we are not specialists, we are likely to be overwhelmed by data rather than find the insights we need.
- Especially in areas where we lack specialist knowledge, we increasingly run the threat of being misled about how our world operates.
- Specialists control their own corners of complex, interconnected systems. Only effective collaboration among many of these specialists leads to desired outcomes. Too frequently, our decisions lead instead to unintended consequences.

The systems that support our specialists remain important: specialization is an important cornerstone of our society today. Yet, the area of greatest challenge has shifted—from supporting specialists to supporting us in areas where we are not specialists. As we live our lives, we feel the crashing waves. The waves of information and confusion from D-Tech may be figurative rather than literal, yet we feel their impact, nonetheless.

Nevertheless, the question of the impact of these *side effects* remains. Are the indirect *side effects* of digital technology truly so malevolent? Is American society stagnating—or is it rushing ahead, despite the hurdles, to delivering benefits? Over the next three chapters, we answer this question, first for our Culture, then Government, and finally Business/Economy.

Are the benefits we obtain from our Culture progressing as that sub-system actively develops and effectively leverages

D-Tech? Or, is American Culture being allowed to passively devolve, with corresponding loss of societal benefit? Let's find out.

CHAPTER 3

THE IMPACT ON AMERICAN CULTURE:

Lack of Neighborhood, and Other
Curses of Digital Technology

In this chapter, we look at Digital Technology (D-Tech) and examine its impact on our Culture. This chapter's foundation comes from my perspective of living on four continents. Those collective experiences helped increase my awareness that "things feel off in the U.S. right now," an impression I make precise and investigate in this book.

D-Tech provides us almost constant access to our loved ones. We celebrate the spouse who proudly sets their cell phone on the table at the start of the meeting because, "My wife may go into labor at any moment." Such innovation is wonderful.

Yet, unfortunately, our culture is failing along too many critical dimensions. In this chapter, we investigate the relationship between D-Tech driven complexity and these cultural failures. For example, through D-Tech, we are hollowing out our relationships. We have the ability to maintain access to more friends and family than at any time in the past. Nonetheless, as guided by our culture, many of us are not applying technology in a way that more powerfully connects us. Instead, we are hollowing out our relationships. As a result, we increasingly face:

> *Deaths of despair (suicides): The population of Americans without a single friend in whom they could confide grew by more than a third between 1984 and 2004. By 2004, more than half of all Americans (53%) felt that they could only confide in a family member; a quarter had no one—not even a family member to confide in (McPherson, Smith-Lovin, and Brashears 2006). Between 1999 and 2014, the suicide rate for middle-aged men increased 43%, and increased an astonishing 63% for middle-aged women (Curtin, Warner, and Hedegaard 2016). Nobel Prize-winning economist Angus Deaton and Princeton Professor Anne Case say, "We might be on the edge of a precipice (Bellini 2018)."*

Our Culture is suffering from the *side effects* (otherwise known as *emergent impacts*) of Digital Technology. We are changing the environment in which our Culture operates. In response, society must solve a new set of coordination problems.

Almost 50 years ago, the D-Tech Revolution began transforming society. The information processing tools blossomed: computers, smartphones, software, artificial intelligence, blockchain, and more. Our world was transformed.

But the impact of D-Tech on our Culture has grown from small ripples to crashing waves. As it has become more powerful, we have created increasing masses of data—which, in many contexts, only D-Tech can reasonably filter. So this technology now determines how we view the world. It overwhelms us, biases us, and depresses us.

Our Cultural experience with D-Tech contrasts with the experience past great societies have had with the technology of their days. As discussed in Chapter 1, as Agricultural Technology was increasingly applied, nomadic societies became settled, and societies in which settlers could thrive had to be developed. In successful societies, leaders reinforced the importance of individual decision-making; citizens protected social discourse that would bring people together.

Today, in contrast, we are drowning in potential. We are not constructing the society to match our current challenges. Instead, D-Tech is perverting our Culture. It stresses our ability to maintain a constructive society. Too many of us are lonely or only partly aware of society's deep challenges. Too many are addicted to drugs or are committing suicide. Our present Culture is like a bear chasing a butterfly: the chase is intense and volatile but also futile. Our Culture is not delivering what our society needs.

However, D-Tech's *indirect* impact on Culture is even more powerful; we call such *indirect (or emergent) impacts* D-Tech's *side effects*. As discussed in Chapter 2, D-Tech has taken control of our information gathering. It's given us masses of information and has connected us with a broad range of contacts. However, it's also confused us. Perhaps it's now too easy to forsake our friend next door for the "perfect partner for this moment," who happens to be in London.

I'm talking with 25-year-old Tauson down the street. He's unemployed and still lives with his mom and dad:

> *"Tauson, who are your friends?" I ask.*
>
> *"I play the online video game Minecraft with Ashish and Jennifer. Akiko is my bud for baseball card trading. I have two brothers I hang out with, and they're a lot of fun," he responds.*
>
> *"Yes, but who's truly your friend?" I continue. "Who do you talk to when you're feeling down? Who helps you figure out what to do when it gets tough to talk to your family? Who helps you plan your future?"*

Tauson pauses for a minute, and then his eyes well up with tears. "I don't know," he responds.

Tauson is struggling:

- He has the perfect partner for every activity imaginable, from baseball card trading to video games.
- He can find more partners, immediately, for any new activity or hobby he imagines.
- Tauson, however, is missing deep friendships: people who may not match perfectly for any specific activity, but are a good match overall. When Tauson had fewer relationships, Tauson would have partnered with such friends on most activities. Through such consistency, these friends would have gained insight into Tauson that today's activity-partners can never match.

Culture today is struggling like Tauson. We have amazing technology that makes it easy to create a virtual connection with Lise in New Zealand or Mike in London. We can find out the score of Peruvian professional soccer matches as easily as we get the results of our daughter's basketball game. We could also text our friend down the street—but that friend no longer has as much of an advantage over our hundreds of other connections. This lack of advantage becomes a barrier to the strong ties with friends and neighbors that can make our physical existence fulfilling.

Over the course of this chapter, we discuss the *second-order side effects* of D-Tech on Culture. This discussion covers the grey-highlighted portion of Figure 3.1. In Chapter 2, we discussed the first-order side effects of D-Tech on our environment: our world is becoming increasingly complex. Over the course of this chapter, we discuss the *second-order side effects* of D-Tech on Culture. This discussion covers the grey-highlighted portion of Figure 3.1. Culture evolves in response to this increasingly complex environment (Arrow 1). How Culture evolves also depends on ongoing evolution of A) actions taken by Government, and B) changes in how Business operates (Arrow 2).

Digital Technology Is Challenging Our Collaboration

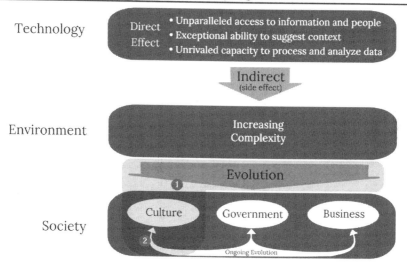

Figure 3.1: American culture has evolved as a result of D-Tech induced complexity, i.e., as a result of D-Tech side effects.

So, what is the most obvious *side effect* of Digital Technology felt by our Culture? D-Tech pervades *every* aspect of our culture. Consider teenagers today: four kids lounge on the porch in late August. School starts in two weeks so guess what they're discussing? School. Books. Girls. Boys. Except, it's silent. They're sitting next to each other but no one is *talking* to anyone. They're *texting* each other from three feet apart!

"R dey really? :("

"You've almost Bin awA 4 6 whol weeks!! THAT'S nsan."
"Time flies yo"[1]

Figure 3.2: Even physical proximity doesn't guarantee direct interaction anymore.

So our Culture is changing. However, change isn't necessarily bad. As Mary Chayko (2016) points out, our technically-savvy young people will soon take over the world. With these youth's ascendance, everyone in society will share similar comfort with today's virtual (as well as physical) culture. In the meantime, as highlighted by Suzanne Kaplan, a multi-generational leadership trainer, "Cross-generational perspectives can enhance productivity. The key is ensuring team members are truly open to exploring new approaches."[2]

There's more, however. What Digital Technology *side effects* (or emergent impacts) are impacting our Culture?

We answer this question here by dividing this chapter into three sections. First, why is there a relationship between D-Tech and changes in our Culture? Here we expand on the increasingly complex environment described in Chapter 2, focusing on the aspects of our environment that most strongly impact our Culture. Second, how has Culture adapted to this complex environment? In short: poorly. We are mangling the critical tools our Culture uses to support society. Third, what is the result of our Culture's adaptation? We are drowning in potential. We are more isolated, more scared, and less secure than in the past. We have relationships, but not the ones we need to solve our most challenging problems.

Across these three sections, we use three elements to discuss our Culture as illustrated in Figure 3.3: Culture forms cultural norms and relationships through social discourse.

Cultural norms are shortcuts in our thinking: the beliefs, attitudes, and expectations we have about possible actions and stigmas. We rarely think about all of our possible options. Instead, we automatically focus on one or a few. Do we eat our hamburgers and fries with our hands and fingers, or do we use a knife and fork?[3] When we see a close, female friend, do we greet her with a firm handshake or a hug?[4] If we invite a friend to dinner, do we expect him to be on time (McMahon 2012)? The answer to each of these questions is a cultural norm: each Culture provides its own, automatic answer to the question.

Cultural norms are diverse.

Both relationships and cultural norms are formed through social discourse. Social discourse is the information and experience framing our lives—plus the systems that support these.

Culture

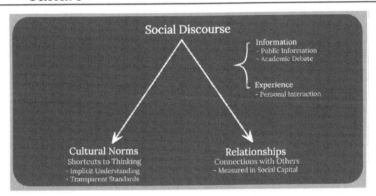

Culture forms Cultural Norms and Relationships through Social Discourse.

Figure 3.3 : As we progress through our life, the social discourse we have impact the relationships we form and our approach to thinking (cultural norms).

Social discourse includes:

- **Public information:** Information publicly available. We spend hours on Facebook to see pictures of little Timmy and to view the gruesome terrorist bombing update.
- **Academic Debate:** Publications and presentations by our academic community. Our academics spend hours in their

labs or pouring over their tomes of Government data in order to publish groundbreaking research.

- **Personal Interaction:** The interactions we have with others throughout our day. These include the superficial discussions about politics with friends over hors d'oeuvres and our in-depth discussion with Tauson about his fears for the future.

These definitions allow us to visualize the impact of technology as it ripples across society. For example, technological development, of the 1940s, 1950s and 1960s brought women into the workforce. Stronger machines and other advances created jobs for people with less physical strength in factories and other workplaces. Many women took those jobs. Women's roles in the family and expectations for themselves began to change. Men and women began to make new life decisions.

In a well-functioning Culture, clear social discourse would help society react constructively to such change. Social discourse would monitor the change announcing successes and problems. How are families becoming more successful? What new problems are being encountered? Such questions would be discussed on the nightly news, parent forums, and at parks.

Debates would be had about the role of women and how legislation and norms about divorce, childcare, and workplace romance should change. Academics would conduct research about the current state and the potential impact of proposed changes.

As social discourse progressed, relationships among those with like interests would form. Cultural norms about women in the workplace and the role of neighbors in caring for children would be revised appropriately. New businesses and government regulations would develop to solve the problems as they arose.

The process would be a human process: there would be tension, quarreling, and regression. Nonetheless, progress overall would be made, and the result would be constructive. In fact, many would argue that such progress was made throughout the 1940s, 1950s, and 1960s.

Over the coming pages, we discuss the actual relationship between D-Tech and Culture. Rather than helping us leverage D-Tech effectively, our Culture is deteriorating. The pressures on how our Cultural systems work are currently proving to be too much.

I judge our culture's success on the wants and needs its citizens satisfy. According to Maslow's Hierarchy of Needs, Figure 3.4, an individual has five levels of needs.

Maslow's Hierarchy of Needs

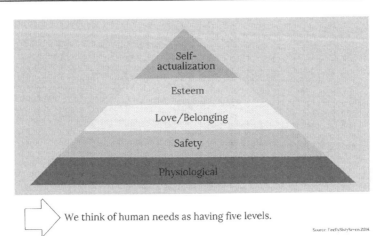

We think of human needs as having five levels.

Source: FireflySixtySeven 2014.

Figure 3.4 : Maslow's hierarchy illustrates human
needs according to an ordered progression.
Credit: FireflySixtySeven (2014).

The base of Maslow's hierarchy represents our physiological needs. In a successful society, children celebrate their parents' stocked refrigerators. One level higher captures our feelings of safety and protection: the goal of every adult. On the third level of Maslow's hierarchy lay our feelings of love and social connection.

In a successful society, we feel the warmth of our friends' love, as well as that of our family and community. When we have high self-esteem, we can actualize ourselves by delivering something we can celebrate. Self-esteem and self-actualization sit atop the hierarchy.

I also judge the impact of our Culture on our ability to work as a group. In this judgment, I include Culture's ability to help us align inconsistencies in our Cultural Norms and across social discourse—so that individuals will be able to benefit.

WHY DOES CULTURE FIND OUR D-TECH ENVIRONMENT CHALLENGING?

D-Tech transformed our environment. In Chapter 2, we introduced three changes to our environment, each of which stresses our Culture. First, those engaged in our Culture must wade through a deep sea of data in Gathering Information we need. Through D-Tech, we can find some information instantly while other things take more time, confuse us, or overwhelm us. Second, we use the information available to us to understand how our world currently operates (Understanding Context)—providing us insight at times, and misleading us at other times. And, third, our Culture must also help us assess complex, inter-related options both more fundamental and more rapidly changing than what our ancestors faced (Analyzing Options).

In particular, D-Tech helps us in our roles as specialists. Where we are not specialists, it is less helpful. Yet, as it has progressed, the size and scope of these challenges have grown. Today, each challenge, independently, can feel overwhelming. In combination? We feel like we're drowning in potential.

Over the coming pages, we will review how each of these challenges is relevant for our Culture.

CULTURAL GATHERING OF INFORMATION

Consider our challenges in Gathering Information for our Cultural decisions. D-Tech is providing us masses of information. Social discourse, our information filter, is struggling. Social discourse is now destructively shifting our world-view and leaving us depressed.

Throughout history, each individual has always been exposed to only a small fraction of the information available. The role of social discourse has been to provide the information we need to obtain our needs and wants. That information we receive must include relevant facts and enough context to understand those facts. Perhaps long ago a warning call would inform us of a bear in the vicinity. A leader's story about courageously defending his family would provide the important context: bears are dangerous.

In many ways, we fail to find the information we need in our D-Tech environment. Certainly, we can find masses of information—instantly! Want an online Bridge partner? 0.4 seconds later, Google has one for us. Want to sell our 19th century Egyptian postal stamp? Again, 0.4 seconds later, Google lists 1.9 million places we can sell it.

However, how about more meaningful information in areas where we are not specialists? We struggle with a parent with Alzheimer's and look to Google's 67 million web pages discussing the disease. We search through thousands of related clinical trials ongoing at any one time. We "scour, hunt, email, call, and text doctors and Ph.Ds in New York, Washington, Baltimore, Phoenix, Los Angeles, Tel Aviv, Stockholm and Paris." We go "a little bit crazy" (Savodnik 2018).

However, the availability of 67 million web pages on the disease obscures the basic truth: we have no cure for Alzheimer's today (Shenk 1998). The 67 million web pages roiling in terms like, "amyloid," "dysphasia," and "subcortical dementia" confuse us without informing us. The thought, "There's so much confusing information, there *must* be an answer in here," tantalizes us. However, we are destined to watch that beloved parent slip away. At some point within the next few years, we are likely to receive a phone call in the middle of the night calling us to the emergency room. We will be saying goodbye. No, that sea of information and the years of searching and going crazy did not help.

Where is the calm, steadying voice that will explain in plain terms what to expect, and that steadying hand that holds ours while we process the situation? That voice—such truth—is more challenging to obtain.

In total, we are increasingly isolated and confused. We can find masses of information. Yet, in too many areas, particularly where we are not already experts, that sea of information confuses. In Chapter 1, we discussed a similar pattern for access to people. D-Tech offers us access to billions of people. Yet, those billions also don't warm us: too frequently, their existence gets in the way of what we need.

Once we have gathered our information, we must come to an understanding of how our world operates (Understanding Context).

CULTURAL UNDERSTANDING OF HOW OUR WORLD OPERATES

All of D-Tech, but especially our social media may appear to be a window. Before D-Tech, we could see our culture through the window of a teen's bedroom. Ricky leans out of his window, the breeze jostling his hair. Three laughing teens stand below pointing up at Ricky. Ricky shouts some news about his friends to the teens on the ground. The teens below respond with a joke. Isn't this what social media is—but with a little added technology?

No. Social media—and, in fact, all of D-Tech— is not a window. Social media is an actively managed, digital media that manipulates our decision and distorts our perspective. The top of a Facebook news feed features postings from only three of our friends. However, we have hundreds of Facebook friends. *Facebook selects which news we see.* We shout out to our friends from the social media window—but whether our friends hear us depends on what the window, itself, wants.

Cameron Marlow, the former head of Facebook's data science team described the potential of Facebook with joy. "For the first time," Marlow said, "we have a microscope that not only lets us examine social behaviour at a very fine level that we've never been able to see before, but allows us to run experiments that millions of users are exposed to (Foer 2017)."

Facebook is constantly experimenting with ways to influence us. Facebook sought to discover whether emotions are contagious, for example (Booth 2014). To conduct this trial, Facebook attempted to manipulate the mental state of its users. For one group, Facebook excised the positive words from the posts in the news feed; for another group, it removed the negative words. Each group, it concluded, wrote posts that echoed the mood of the posts it had reworded. This study was roundly condemned as invasive, but it is not so unusual. As one member of Facebook's data science team confessed: "Anyone on that team could run a test. They're always trying to alter people's behavior (Sherwell 2014)." Facebook has claimed that it may be the primary reason for the increase in voter turnout from 2006 to 2010 (Bond et al. 2012). Facebook can predict users' race, sexual orientation, relationship status, and drug use on the basis of their "likes" alone (Foer 2017). Such information is, of course, valuable for identifying the best way to influence us.

Social media also indirectly determines the emotion with which we face the world. It frequently depresses us. How is that? So much of what we see seems so upbeat. A middle-aged father writes about his pregnant wife and growing family, "Three years ago, we were 2 ½ strong. Now, 3 ½ strong. Age 45 will be busy." How does something seemingly so benign depress us?

Social media's distorting filter depresses us. Our friends post the positive aspects of their lives. We don't see the "happy father" with massive bags under his eyes, appearing 15 minutes late for a work meeting smelling like diaper juice. We only see the smiles.

Meanwhile, in our real life? Hmm. My baby just woke us up at 2:00 a.m. and then threw up all over my wife's work dress that was hanging off the ironing board. The dress had been lying the-re for two-and-a-half weeks because we just couldn't find time to iron. My Aunt Jenny got fired and her 23-year-old daughter is still living at home. If cousin Ricky can kick his opium addiction, he'll probably graduate from junior college; it's a funeral or a graduation ceremony. We know the odds and we're praying.

Is that view of life exaggerated? Sure. However, it reflects the real outcome: social media makes us envious (Appel, Gerlach, and Crusius 2016). Facebook users present the most positive aspects of their lives. Our own, real lives, simply don't measu-re up. Meanwhile, there are 202 million Facebook users in the United States (Statista 2018), and each spends an average of 50 minutes a day on the site (Stewart 2016). The more time a user spends on Facebook, the more depressed they are (Ibid.).

Through D-Tech, we also frame our discussion on sexuality. Four years ago, Facebook allowed members to define themselves as either of 2 sexes. Today, Facebook offers 56 options including gender fluid, pangender, and trans male (Oremus 2014). Because most Americans are on Facebook, these gender options matter: they define the dimensions with which we discuss gender and sexuality.

Such D-Tech influence arises from many origins outside social media, as well. One sophisticated D-Tech method completed the comparison, "Man is to computer programmer as woman is to homemaker (Bolukbasi et al. 2016)." As discussed in Chapter 2, another only highlighted men when a user searched for "CEO."

Any decision about how to represent women CEOs, Jews, people of color, or white supremacists impacts our culture. Any decision can be right or wrong—or both right *and* wrong, depending on the person's perspective. Programmers use seemingly clinical mathematics to make these decisions. However, these decisions are not clinical—they determine how our society communicates. It may seem these decisions written in a programming language can be neutral. In fact, they can't.

A number of organizations are studying such impacts. These groups include an Ethics and Society Impact team organized by Google's Deep Mind (Condliffe 2017), an industry-sponsored group called The Partnership on AI (Artificial Intelligence) (Knight 2016), academic-sponsored groups at Carnegie Mellon University (Russell 2016), and the AI Now Institute (n.d.). Despite the study, however, these challenges are currently not *under control*.

So, how shall we summarize what's happening? Increasingly, we rely on D-Tech for a perspective on how our world works. This D-Tech filter, therefore, structures our world-view. D-Tech relies on clinically precise, mathematical equations—but that filtering is not benign. D-Tech is frequently selecting what information to deliver to us, and how it is presented—and the D-Tech selections are impacting our perceptions of the world around us and how it operates. Once we understand how our world is currently operating, we must decide what we want to do, by Analyzing Options.

CULTURAL ANALYSIS OF OUR OPTIONS

As our D-Tech world has become more complex, analyzing which option best suits us is more challenging. Our social discourse plays a key role in assisting with such analysis: deciding which of our ever-expanding options we want to explore. Such questions of what we want exist at every level of Maslow's Hierarchy of Needs.

Let's begin with the bottom level of Maslow's hierarchy: physiological needs. A fundamental physiological question is what to do with those people who have fared poorly in our D-Tech economy. As illustrated in Figure 3.5 (Piketty, Saez, and Zucman 2016, 46, figure 5), D-Tech has been more generous to our wealthy citizens than to our middle and lower classes. The clear line illustrates the faltering earning by the poor: from 1962 through 1976, the bottom-half of wage earners made 20% of our total American income. Today, this half of the population earns only 12% of our total income. The shaded line illustrates the rising earning of the rich: In 1966, the top 1% of income earners made, in total, 12% of the total income. Today, this 1% of the population earns over 20% of our total income. What is fair and what are the implications of any (un)fairness? And, do we want to do anything about it? Those are questions for our social discourse.

How about protecting our safety, the second level of Maslow's hierarchy? What is the definition of violence in a D-Tech environment? Through current D-Tech communication, we can send messages to someone's phones and computers. In one extreme, we already have a legal case of a girlfriend using text messaging to convince her boyfriend to commit suicide (Sanchez, Lance, and Levenson 2017). And, the next phase is just around the corner. We are implanting microchips in our brains to monitor our brainwaves (Piore 2017) and allowing people to operate a keyboard directly with their brain functions (Thompson, Carra, and Nicolelis 2013). Elon Musk is funding research to enhance these direct brain-computer connections (Stibel 2017). And, rats have already played real-world games in which they communicate with each other solely by computer chips implanted in their brains (Pais-Vieira et al. 2013). When we can almost directly plant information in another human's brain, what constitutes safe interaction—and what is violence?

Pre-tax National Income Share: top 1% vs bottom 50%

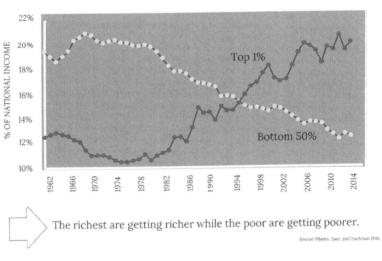

The richest are getting richer while the poor are getting poorer.

Source: Piketty, Saez, and Zuchman 2016

Figure 3.5: Over time, the share of income earned by the wealthiest Americans has grown, while the income of the poorest Americans has shrunk.

Our social connections, the third level of Maslow's hierarchy, are similarly being re-defined. Do we want to accept machines as social companions or mates for our elderly? Our children? Or ourselves? [5]

What determines our self-esteem is also being re-defined. We used to compare our bodies and minds to our neighbors. Today, we must compete with the most beautiful and brightest in the world—we see and hear them constantly. And, of course, we have our ever-present concern about self-esteem in a world that could become jobless. As a society, we have questions to answer beyond what our parents had to consider (Thompson, Derek 2015). We must ask more from our social discourse than ever before.

Such broad, societal discussions are important. However, we also must answer more day-to-day questions. We cannot think deeply about every decision we must make. We must automate our response to many more questions than we did in the past. Cultural norms are the tools society uses to help us automate such decisions.

We look to our own experiences and what our neighbors are doing as we decide how to raise our kids, for example. Often, we simply copy the decisions of Bob, Sue, and Carlos whom we meet at the park. They share stories about how they raise their children, and we hope these stories reflect good decisions. To provide meaningful assistance, we need these cultural norms to be more nuanced than ever: we need help with more than just deciding when to give a cell phone to a child. We also need help with all the follow-up questions about how we let the phone be used. Once my son has a cell phone, for example, should I monitor his Facebook account? Should I let him choose which apps to download? Should I let him surf the Internet without supervision? And are the actions I take for one child also appropriate for my other children?

The world is changing fast. We would like our cultural norms to keep pace. Parts of the Internet that were clean and safe become rapidly polluted when someone finds an automated way to make them so. Then other parts become clean. The speed of change is breathtaking. That's challenging.

SUMMARY

In short, D-Tech has created a challenging environment for our Culture. In Gathering Information, (i.e., finding the data we need), we must apply D-Tech's filters. While we have unparalleled access to people and information, that same access to masses of information is likely to confuse, especially where we are not an expert, or are deeply and emotionally involved in the outcome. Meanwhile, the billion people don't provide us comfort—our masses of weak ties are hurdles to creating deep connection.

We use our D-Tech information to understand how our world is operating—this information structures our world-view. Yet, D-Tech and its mathematical equations is not a benign presence. D-Tech and its programming selects our bias—and, too frequently, that bias is destructive, especially emotionally.

To Analyze Our Options, we need more help with questions big and small. Our social discourse must help prompt an increasing number of important questions, and help us answer these questions constructively. This social discourse must allow society to answer immense questions about our future—questions that will require expertise of many types to answer.

Social discourse must also help us automate decisions not worth thinking about. Today, we are overwhelmed with such decisions.

So, these Cultural challenges are endemic to our D-Tech environment. How has our Culture evolved in response? Have we strengthened our Culture to answer such challenges? That is our next topic:

HOW HAS CULTURE ADAPTED TO OUR D-TECH ENVIRONMENT?

Our Culture is letting us drift on an increasingly stormy sea. Yes, the salty spray is just beginning to sting our faces. The din of thunder remains far away as does the darkest of the storm clouds. Yet, we're ignoring these calls to stabilize our anchors.

Our anchors are our cultural norms and relationships, and our environment provides the opportunity to shift them. Through our social discourse, our environment is translated into the information and experiences that shift our cultural norms and relationships. There are two ways in which this shifting happens:

- Our public information can act like the powerful wind that fills a boat's sails. We read articles about social disintegration, and we become less trusting. We hear about a hero teacher saving children from being shot in our school, and we value our teachers more highly. Over the coming pages, we will discuss how, rather than blowing our cultural norms and relationships in a clear direction, today our social discourse is simply blowing us hard . . . in every direction.

- Our own experiences can act like the currents that tug at the anchor's line. A constant drumbeat of messages from our leaders about the importance of children, for example, could reinforce such a cultural norm. At home, or church, or at work, we'd likely hear and see such consistency. We will also discuss in the coming pages how our experiences do not share a consistent theme; they are pulling us in many directions, not leading us confidently in one.

PUBLIC INFORMATION

What's happening with our public information? Our public information is distorted: it's insulating, emotionally exhausting, and antagonizing. Meanwhile, our popular information supports our further corrupting of that information. Our popular news is supported by an industry insufficiently motivated to provide quality content and a government that doesn't support truth. Meanwhile, too many academics are chasing clouds. Rather than providing a coherent view of our world—instead of providing each of us the information we need to stabilize our world—our social discourse is destabilizing us.

We discuss our public information in terms of popular information (such as television news) and academic debate.

How has the system delivering our popular information responded to our D-Tech environment?

It killed Walter Cronkite.

No. Our popular information didn't literally kill anyone. Our popular information killed the memory of the man symbolizing coherent, high-quality, popular information. As we will discuss over the coming pages, our news industry is creating increasingly flawed stories. These stories are then stretched, torn, and filled with mud online. Relatively unimportant stories are delivered to an increasingly insulated, emotionally exhausted populace.

Walter Cronkite was a news anchor before Digital Technology. He ended each news show with the same line throughout the 1960s and 70s: "What kind of a day was it? A day like all days, filled with those events that alter and illuminate our times. And you were there (Vance, n.d.)."

As millions of Americans sat in their shag-carpeted living rooms watching, we felt informed when Walter Cronkite said that. Many families watched the news together. Dad would whisper at the screen, "Stupid commies!" Mom would yell, "Ugly capitalists!" Families came together, at least to be in the same room.

Walter Cronkite was "the most trusted man in America." He was more trusted than the president, vice president, and any senator. Americans simply believed that "Walter Cronkite would not knowingly deceive them (Clark 2006)." While Cronkite may be exemplary; he simply mirrored the sacred, journalistic Code of Ethics. That code states, "Journalists should be courageous in gathering, reporting, and interpreting information... Ethical journalists should be accurate and fair (Society of Professional Journalists 2014)."

How did we kill Walter Cronkite? A combination of social media, an unsupported news industry, and incongruous regulation.

HOW OUR CULTURE KILLED WALTER CRONKITE

Walter Cronkite had a news program. Once Walter Cronkite had your attention, he generally "had you for the evening." Over the course of that evening, he could provide a coherent context for each story being presented (Vance, n.d.).

Cronkite's show wasn't perfect. Perhaps yellow journalism,[6] developed over one hundred years earlier by William Hearst and Joseph Pulitzer, had a place on Walter's show. However, Cronkite showed the same yellow journalism to everyone. Just like the "Spanish Treachery" newspaper title everyone saw in 1898 (later proved false) that helped trigger the Spanish-American war (Great Projects Film Company 1999), everyone received a complete, and reasonably similar, view of the world.

Figure 3.6: Families used to come together more frequently in front of the television. Being together didn't necessarily guarantee a meeting of the minds, however.

Fewer people watch today's version of Walter Cronkite, just like we no longer call out our window to our friends. For many of us, no one like Walter Cronkite provides commentary on our news. Today, increasingly, it is the blogger or social media poster who sees the article and reads the posted comment by someone else. The blogger or poster then writes his or her interpretation of the facts. And suddenly this becomes news.

We also increasingly use D-Tech to filter our information for us—and D-Tech has our attention for two-minute-bursts: that's the length of the average Facebook interaction (Taylor 2013). The businesses operating that technology must make every second count. First and foremost, our social media is trying to keep us there for that full two minutes. Social media provides a constant triggering throughout that time to maintain our attention.

So, what happens? Social media maintains our attention by transferring us from one emotional hook to the next. As we discuss over the coming pages, we open our social media window—and we hear insulating, exhausting, and superficially debated topics. In our two-minute sessions, there's little time for social media to include quality information; important topics are shunted aside. Our attention is fleeting and it must be captured.

What does use of social-media look like? It's 10:30 p.m. The kids are asleep. It's just you and your spouse—and your cell phone. Your customized social media news feed begs for your attention: just one more article about the third-string player on your sports team. Just one more article about the bed and breakfast your high school classmate just opened in Bogota, Colombia. You'll regret the decision later, but reading about that third-string player is likely to beat out the 10 minutes with your spouse that should be ending your day.

The loss of such interaction is harmful. The more time we spend on social media, the more we regret it, and the more depressed we feel (Time Well Spent, n.d.). "Use of Facebook [is] negatively associated with overall well-being . . . Liking others' content and clicking links . . . reduce[s] self-reported physical health, mental health, and life satisfaction (Shakya and Christakis 2017)." Social media use gets in the way of the most satisfying interaction—face-to-face relationships (Leung and Lee 2005).

How does social media such as Facebook get in the way of face-to-face relationships? The average smartphone user checks Facebook 14 times a day (Taylor 2013). In the extreme, we can even become addicted (Christakis 2010).

Too frequently, D-Tech's controllers target insulating, emotionally exhausting, and antagonizing content. These targets pander to our desires. Pandering provides these controllers with a few extra seconds of our using their service.

We are insulated because we want to see ourselves as right. This desire may contrast with our view of ourselves as truth-seekers: righteous adventurers out to understand our environment. Unfortunately, the reality is less noble. Mathematicians say we have confirmation bias; psychologists say we suffer *cognitive dissonance* ("Cognitive Dissonance" 2018). Both of these statements mean we like to think we're right—and feel uncomfortable when we're wrong.

Yes, superficially, it means that we are cocooned in a bubble. However, the implications for these bubbles are even more insidious. Such bubbles simplify the process of making false information accepted fact for specific sub-cultures.

Consider, for example, the Internet treatment of the 2015 U.S. military training exercise called Jade Helm 15 (Lamothe 2015). This exercise has been described as follows: "Isn't it obvious? If Operation Jade Helm were happening in any other country, it would be immediately labeled a military drill for martial law (Adams 2015)." Jade Helm 15 is also used to support the statement, "The federal government and the U.S. Army have been readying for domestic unrest for decades, and preparations are accelerating exponentially (Watson and Jones 2015)."

On social media and the Internet, such perspectives tend to be initially shared with those most susceptible to believing this information. These believers share the information with others, and discuss how it must be true. Slowly, such information pervades everyone in a particular, online community until that belief is fully accepted as proven fact. Meanwhile, each person in the online community has only limited newsfeeds to look at; the filter bubble that has formed around each individual doesn't offer up any alternate views of reality. Everyone is now trapped in an echo-chamber, "...there are so many roups of people online—particularly on Facebook—that steadfastly believe information that is demonstrably nonsensical (Andrews 2016)." As Giuseppe Granieri (2014) describes, "this culture tends to reinforce more than it challenges one's existing preferences or ways of doing things."

If we're like most people, our newsfeed is increasingly sheltered. We are focusing our attention on information that reconfirms—rather than challenges—our worldviews. As a result, we are experiencing an increasingly bifurcated world of strong believers and dis-believers—and a fact-base that has many, many flaws (NPR Staff [2016], Kim [2016], Leitch [2014], Chamorro-Premuzic [2014], and Jones [n.d.]). In short, Digital Technology insulates us against ever realizing that we are wrong by insulating us from competing views. It shields us—even if those views may be more valid than our own.

Moreover, we are emotionally exhausted because D-Tech wants us to be connected to it—and creating that connection exhausts us. We need the information that will help us understand our world and make the decisions that will best help us live the life we want. Social media, on the other hand, simply wants our attention. And, to get our attention in an informationally smog-filled world, D-Tech is resorting to "sticky content" more and more often.

What's sticky content? It's content certain to trigger our emotions. For example, my wife buys children's books and toys online and reads about how to nurture a rebellious toddler. Facebook rewards her with horrible stories of childhood pain: "Vile dad beat and raped his children for 22 years and battered his wife (Wilkinson 2017)." "5-year-old dragged across yard by pit bull (Associated Press 2017)." "A routine day's killings: investigating children's gun deaths (Younge 2017)."

Even science coverage is focused on "triggering" us. "Drinking a glass of red wine is equivalent to an hour in the gym (Sitch 2017)." "Eating chocolate helps you lose weight (Cohen 2015)." These headlines will get many busy adults to click through and read—even though they are false.

Social media even invites us into battle to engage us for a few seconds more. Such social battle is based on emotions such as outrage. Rarely is it based on logical debate. Battle integrates yellow journalism with commenting, made possible by D-Tech. Battle fits neatly into two-minute chunks of time, while quality debate takes serious, ongoing thought.

What do such battles look like?

The first grenade was thrown at 2:00 a.m. by a supporter of victim's rights: "This is horrible! How could the attacker have done that?" Eight hours later, a defendant's rights advocate lobs back, "You're disgusting! The defendant deserves a trial!" And the war is on. Battle lines are drawn, and both sides get to show the extremes to which they will go for their cause.

What event catalyzed this online battle? In 2006, a black stripper, dancer, and escort alleged that three white Duke University lacrosse players raped her. The acting District Attorney may have even implied the act was a hate crime ("Duke Lacrosse Case" 2018). Here was a horrific claim with virtually no supporting evidence. The Internet was engaged. Two more horrific rape stories also supported by little evidence became internet sensations soon after ("Tawana Brawley Rape Allegations" 2018; "A Rape on Campus" 2018). Major internet battles ensued between supporters of victim's rights and those supporting rights of the accused.

All three viral, rape sensations later turned out to be false. That falseness is striking—because most rape allegations across society turn out to be true (Newman 2017).

In fact, that falseness gives us a clue as to what happened. The key was the ability to easily draw superficial, internet battle lines. A rape that clearly happened? That disgusts everyone; we all agree. When such crimes are reported, we are all upset, and the event is quickly forgotten. There's no battle. However, a particularly horrid act—that may not have happened? Those supporting victim's rights are almost certainly drawn into a battle with those supporting the accused. The energy from the battle draws in more of us on the internet—and the interest drags on (Alexander 2014).

A constructive debate about how to improve society? That would require a deep understanding of details, policies, and actions that few of us have. It also requires more than two minutes.

Others have learned to create similar battles: draw in an audience by creating a dubious scenario on the outer fringes of what might be acceptable. Then watch as supporters on both sides battle those opposed. There is almost a guarantee of a viral reaction. For example, the water in Flint, Michigan was contaminated by lead. The animal rights group PETA offered to pay the water bills for people living in Flint, but only for those families who agreed to stop eating meat. An online battle ensued. Perhaps PETA was taking advantage of the wounded, "extorting" those in need. Or, perhaps PETA was justified in using any means available to advance its cause. Each side argued vehemently how "right" they were. In the process, PETA received attention, which was, no doubt, a campaign goal.

In such cases, we are drawn into battles that tug on our emotions and show our team's colors. We are not provided information that will spur the social discourse we need to make important decisions.

When there's no trivial battle, the topic itself is often tangential to the truly important questions. What, for example, created a massive stir in social media in 2017? Transgender bathrooms. Transgender bathroom use is a serious issue for those involved, especially for the children involved. However, only an estimated 1 in 200 Americans are transgender (Flores et al. 2016). This topic's importance in social discourse is outsized relative to its social impact.

Meanwhile, between 10 and 22 in 200 Americans have ADHD (Attention Deficit Hyperactivity Disorder)(Holland and Riley 2014). Children with ADHD, as well as their classmates, can find it difficult to learn, and parents of ADHD children find their marriages strained (Harpin 2005). Despite the larger social impact of ADHD, it received only a fraction of the attention garnered by transgender bathrooms.

What was most distressing about the transgender bathroom debate? That debate failed to spur development of the most impactful (and noncontroversial) support for trans children. For many people, realizing your son or daughter is transgender (or gay) is exhausting. Challenging parenting topic after challenging parenting topic must be handled:

Figure 3.7: Information about which we can argue gathers
attention, whether or not it deserves that attention.

- How to speak with the trans child.
- How to parent any siblings who are not trans when the trans child is receiving seemingly all of the attention.
- How to stabilize a relationship between parents when one parent is more accepting of the trans child than the other.

- How to discuss the trans child's needs with schools and activity programs.
- And more.

Despite the massive societal focus on "trans," there remains no structured program to help parents of trans children manage such highest-impact challenges. Two women I spoke with recently are launching such a parent-focused program—and I hope they rock the world.

Did Walter Cronkite include low-quality, eye-catching content in his programs? Sure. However, such fluff pieces were planned elements of 60-minute programs. Today, our social media programmers are focusing on keeping us for two minutes. Providing both fluff and more meaningful content all within two minutes is a far more challenging, if not impossible, task.

In short, our Culture has acted to further expand the destructive ripples of Digital Technology. We are drawn toward sticky outrage. It keeps us there, engrossed in topics that depress us (Neo 2017). We share this depressing content with others. We battle a dehumanized army of foes with whom we disagree on trivial topics, and, in the process, further bias the already-biased information we share. We confirm our own biases. All in two minutes! We learn little that allows us to function more effectively in society. And we walk away exhausted. "These days, we can usually get what we want very quickly. But we can't always get what we need to lead fulfilling lives or to construct a society we're satisfied with (Dionne 2014)."

D-Tech and social media do strengthen some physical relationships (Oh, Ozkaya, and La Rose 2014). And, the creators of D-Tech are human. They recognize that they have an overwhelming impact on today's society—and that we know that as well. Facebook recently modified its approach, as it often does. CEO Mark Zuckerberg said, "Protecting our community is more important than maximizing our profits... [Facebook will] measure itself on the number of interactions that people have on the platform and off the platform that people report to us as meaningful (Constine 2018b)."

The change Mark Zuckerberg called for may reduce the negative impacts we've just described. Withal, the impact will depend on exactly what Facebook—and other purveyors of D-Tech—actually do (Constine 2018a). Will Facebook truly measure itself solely on meaningful interaction—or will profitability remain important? What does Facebook consider a meaningful interaction? Can an interaction be meaningful if it doesn't engage us with differing opinions, or if any engagement is only superficial?

In addition, Facebook is an inherently challenging medium within which to target socially meaningful interactions: two-minute interactions are *fast*. Facebook is reliant on engaging with us for every second. Facebook's change may also make some challenges worse. For example, with Facebook's recent change in algorithm, we'll see less news (Constine 2018a). Perhaps, however, we'll simply see less low-quality news. The impact of D-Tech on news quality is the topic in the next section.

HOW BUSINESS KILLED WALTER CRONKITE

Walter Cronkite and the firms that employed him invested in his reputation. Walter had to put in the effort to ensure his news coverage was solid. Perhaps there were nights he got less sleep or days he missed his kid's baseball game. Meanwhile, his firm invested in fact checkers, journalists, and databases needed to provide quality, public news. His firm made this investment in quality because it was rewarded with profit.

Today, we don't trust our news. Less than one in three Americans has even a fair level of confidence that our news media will "report the news fully, accurately, and fairly (Swift 2016)." Unfortunately, it appears the lack of trust has been earned: Erin Gray (2016), for example, describes how news is sloppily reported. Fact checking is done poorly (Lim 2017). Journalists rush to get "a story" within the news cycle without the budgets and fact-checkers to do this consistently well.

Journalism used to be a high-revenue business. Entice the crowds by developing a quality reputation—and there was profit to be made. The source of that profit: "Buy Brylcreem Hair Cream, available cheaply at Woolworths." The source was advertising.

Revenues from news advertising are a small fraction of what they once were. Craigslist essentially eliminated the revenues that came from personal advertisements (Weiss 2006). The Internet has massacred ad revenues linked to newspaper circulation (Economist 2006). The option to get "free news" from the local blogger has reduced the profit potential for news programs. Journalism industry stalwarts who once profited from strong reputations earn little money today. *The Washington Post*, for example, is essentially earning no money. It did report a surprise profit in 2016; however, that profit was most likely small and came after years of losses (Bedard 2016). Meanwhile, advertising revenue at *The New York Times* has fallen more than 50% from peak levels (Polgreen 2016). Yes, *The Times'* revenues recently rebounded modestly, but that is from exceedingly low levels (Ember 2017).

The news industry has responded in three ways to such pressures on profit—all decreasing the factual quality of our public news. First, they reduced the direct investment in ensuring news quality. No news organization can afford to pay large sums to develop or maintain a strong reputation for journalistic ethics.

The eventual reward in profits simply isn't there anymore. These tighter budgets are not up to today's faster news cycles, increased complexity, and increased importance of nuance. For many, many public new sources, there simply isn't the workforce available to provide quality, fact-checked stories.

Second, firms in our public, news markets are embracing Government and other biased news sources. Consider close-hold embargoes (Seife 2016):

> It was a Faustian bargain—and it certainly made editors at National Public Radio squirm.
>
> The deal was this: NPR, along with a select group of media outlets, would get a briefing about an upcoming announcement by the U.S. Food and Drug Administration a day before anyone else. But in exchange for the scoop, NPR would have to abandon its reportorial independence. The FDA would dictate whom NPR's reporter could and couldn't interview.

NPR was not allowed to get non-Government responses to the story before running it. In short, the U.S. Government would tell NPR what to say. In addition to NPR, CBS, NBC, CNN, The New York Times, and the The Washington Post all accepted this 2014 deal from the U.S. Food and Drug Administration.

In a close-hold embargo, journalists agree to make no efforts to check with other experts about a story until a preset time. At that pre-set time, they—and all of their competitors—are allowed to publish the story simultaneously. With the criticality of being the first to release a story, most stories under close-hold embargoes are published before any fact-checking can be done or opposing views obtained.

If you are not worried about your reputation, close-hold embargoes are attractive. You spend no time or resources checking the story—because you aren't allowed. Meanwhile, you're confident your competitors also won't do any checking—because they're not allowed. At the pre-set time, you and your competitors simultaneously release the story—so there's no need to even rush to get the story out. Once the story is out, the news cycle moves on. Incentives for journalists to go back and do the proper research on such an "old news" story are unlikely. This agreement may be unethical, but it sure is cheap! The public, meanwhile, is cheated out of public news independent of the news source.

How prevalent *are* news embargoes? Surprisingly common (Seife 2016):

> *A surprisingly large proportion of science and health stories are the product of embargoes. Most of the major science journals offer reporters advance copies of upcoming articles—and the contact information of the authors—in return for agreeing not to run with the story until the embargo expires.*
>
> *Despite the difficulty of measuring the use of close-hold embargoes, Oransky and Kiernan and other embargo observers agree that close-hold embargoes—and other variations of the embargo used to tighten control over the press—appear to be on the rise. And close-hold embargoes have been cropping up in other fields of journalism, such as business journalism as well. "More and more sources, including government sources, but also corporate sources, are interested in controlling the message, and this is one of the ways they're trying to do it," says the New York Times' Sullivan.*

U.S. Government-produced propaganda is receiving a similarly warm welcome from the news industry today. Propaganda certainly has a proud history in the U.S., as then-President George W. Bush (2005, 5) described:

The use of government money to produce stories about the government that wind up being aired with no disclosure that they were produced by the government ... has been a longstanding practice of the federal government.

Remember "Rosie the Riveter" (Miller 1942) from World War II? The concept of Rosie was inspired U.S. Government propaganda:

Figure 3.8: Rosie the Riveter illustrates positive government propaganda, developed under the watchful eye of a relatively strong news industry.

Shortly after World War II, the U.S. Government got out of the propaganda business, however. According to the 1948 "Smith-Mundt Act (the U.S. Information and Exchange Act), no Government propaganda was to be released within the borders of the United States.[7]

Despite the Smith-Mundt Act, the U.S. Government began expanding its delivery of propaganda during the D-Tech era. In 2005, for example, a reporter for Florida television station WBHQ seemed to corroborate then-President George Bush's foreign policy. The report described how "Afghan women, once barred from schools and jobs, were at last emerging from their burkas, taking up jobs as seamstresses and bakers, sending daughters off to new schools, receiving decent medical care for the first time and even participating in a fledgling democracy."

In short, "American intervention abroad was spreading freedom, improving lives, and winning friends (Barstow and Stein 2005)."

Private news journalist Tish Clark narrated that report. However, U.S. Department of State contractors conducted the interviews, selected the quotes, and even wrote the narration that Ms. Clark presented (with limited edits). Despite the U.S. Government essentially authoring that report, neither the public audience, nor even the narrator Ms. Clark herself, realized the report had originated with the State Department.

Such reports are now common. Federal agencies have been commissioning video news releases since at least the first Clinton administration (1993-1996). President George W. Bush's administration spent $254 million on a public relations contract during its first four years in office (2001-2004). Even by 2005, at least 20 federal agencies including the Defense Department and the Census Bureau had made and distributed hundreds of television news segments. According to one video news release company's pitch to potential clients, "90 percent of TV newsrooms now rely on video news releases."

D-Tech didn't force Government to reinstate propaganda development. However, D-Tech did create the news industry environment that gulps down the (freely available) propaganda. Television station WCIA's news department, for example, was cut from 39 to 37 employees at the same time as the content those employees needed to fill went from 7 hours to 8. It shouldn't be surprising that, in a single three-month period, WCIA ran 26 segments produced by the U.S. Government's Agriculture Department alone (Barstow and Stein 2005).

The U.S. Government, meanwhile, isn't complaining about the "unfiltered message, delivered in the guise of traditional reporting (Ibid.)." In fact, the U.S. Government pays distributors and networks to pass on the content.

It's possible such propaganda production was technically illegal during the late 20th and early twenty-first century.[8] In 2013, however, the U.S. Government eliminated that last vestige of its post-World War II decision to stop producing propaganda. In that year, the Government eliminated the Smith-Mundt Act limits on propaganda as part of the National Defense Authorization Act (*Public Law No. 112-239 [01/02/2013]*). The Government then funded the creation of propaganda for our national audience in the following year's National Defense Authorization Act.

Such propaganda does not necessarily weaken social discourse. Arguably, Rosie the Riveter was an inspiration that helped win World War II. Government propaganda is a tool like any other.

However, we wouldn't put a gun in a room full of four-year-olds: a gun can only be used safely by well-trained adults. Similarly, the threat of Government propaganda depends on the state of the independent news sources. Today, our public news is weakened, as we've been discussing. Using Government propaganda in a nation with weak independent news is putting a gun in a room full of four-year-olds. It threatens Culture's ability to have a strong, healthy social discourse.

Third, business delivering less diverse sources has diminished the quality of our public news. Today, five conglomerates manage 90% of what most Americans read, watch, and listen to (Sommer 2014). As of 2015, these five conglomerates had reduced journalist employment by more than 40% from its 2006 peak (Edmonds 2015). The ability for D-Tech to improve journalists' efficiency certainly played a role in this reduction; however, the reduced reward for quality—and reduced competition to be perceived as high-quality—also likely played a material role.

As the number of journalists decreased, their viewpoints became less diverse. Figure 3.9 illustrates how journalists are increasingly living in areas that have consistent political preferences. For the 2008 Presidential Election, 39% of journalists lived in counties that supported the presidential candidate of one party. The other 61% lived in counties supporting the other presidential candidate. That's not an equal split, but it's not too bad.

By 2016, that split was considerably less diverse: a full 72% of American journalists lived in counties supporting the same major presidential candidate; only 28% lived in counties that supported the other (Doherty and Shafer 2017). In fact, in 2016, 51% of journalists lived in counties where one presidential candidate was supported by at least two out of every three people who voted. Almost everything such journalists heard in 2016 from their neighbors was most likely in support of that one candidate and derogatory toward the other one. Maintaining even-handed news coverage must have been exceedingly challenging for those journalists. More likely, we received public news reflecting the bias of the counties where the journalists lived rather than reflecting national sentiment.

In short, our markets have harmed the factual quality of our public news. D-Tech has changed our environment. Our world is now more challenging to report neutrally. Business dynamics have increased this challenge by reducing the profit that news organizations can obtain from high-quality news. Businesses have responded. They've reduced their investment in quality; they've welcomed cost-saving albeit quality-harming, close-hold embargoes; and, they've concentrated journalist employment in regions that share similar political perspectives.

We can't blame the news industry for all quality issues. Social media, itself, also plays a key role. We increasingly source public news through social media (Anderson and Caumont 2014). "Findings" are "reposted." Re-posters add their own, personalized biases and misrepresentations. "Bloggers transform the message to fit their own identities, as well as their audiences and networks (Vaick 2012)." Information is continually reposted, and details continue to morph with each repost as re-posters add their own, unique spin.

Journalist Diversity

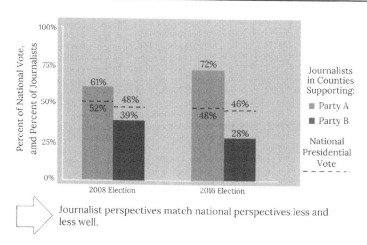

Journalist perspectives match national perspectives less and less well.

Figure 3.9: An increasing share of journalists are coming from areas tending to support only one of the political parties. As a result, our journalists' perspectives are becoming less diverse (Doherty and Shafer 2017).

Is it possible to exaggerate the negative impacts? Yes. Technology is increasing the efficiency of journalistic efforts. Internet searches certainly help with fact-checking. Writers can now create computer-generated articles with the push of a button in some situations. (Examples include stock market summaries, Keohane 2017). Meanwhile, bloggers and other unpaid journalists frequently also provide many high-quality nuggets that are only possible because of D-Tech. Nonetheless, the primary impact is clear: pressures originating in D-Tech have rippled through

Business and are now challenging our public news. Such news is central to modern, social discourse.

Facebook is promising to provide us with less news. The indirect impacts of that change will be interesting to see. Perhaps it will force more of us to pay news providers directly. That could help improve our news quality. However, it would be helpful if our Government supported quality news. Today, it's not. That's our next topic.

HOW GOVERNMENT KILLED WALTER CRONKITE

Government played a key role in supporting constructive Agricultural Revolution Culture thousands of years ago. Hammurabi, for example, created key regulations supporting constructive social discourse.

Today? Our Government is acting destructively toward our Culture's Social Discourse in all too many key instances. Perhaps our Government became jealous of Walter Cronkite. Perhaps now the Government doesn't want us to trust what we see on the internet. Whatever the cause, Government is currently supporting destructive information-sharing: the U.S. Government has developed a set of rules and precedents that have weakened the factual quality of our public news.

As we'll discuss further in the next chapter, Government has lost its ability to lead as we've increasingly implemented D-Tech. Complexity, and the related unintended consequences, have swamped our Government's ability to set and execute a clear vision for society. Special interests have stepped in and helped create a confusing web of regulations and judicial precedents. Particular elements of this confusing web are destructive to our social discourse.

Consider the case, for example, of Yelp. Yelp is a website that lists local businesses with user reviews (Yelp, n.d.). Positive user reviews on Yelp can bring in customers: "The avocado dumpling was awesome!" Meanwhile, "The catfish was disappointing, bland, and mushy" can drive customers away. At the same time, businesses featured on Yelp can pay for advertising help, and this is where the role of Government is important.

Cats and Dogs Veterinary Hospital, for example, sued Yelp for misrepresenting their business (*Levitt v. Yelp! Inc., at 7*):

> *Cats and Dogs Veterinary Hospital contacted Yelp to request removal of a negative review by "Chris R," where that review was inconsistent with Yelp's policies. That review was removed.*
>
> *Several days later, a new negative review of Cats and Dogs, from "Kay K," appeared.*
>
> *Soon after the appearance of these negative reviews [Cats and Dogs] began receiving frequent, high-pressure calls from Yelp sales representatives, who promised to manipulate the Cats and Dogs listing page, "hiding negative reviews" in exchange for [Cats and Dogs] purchased advertising.*

The judge in the case ruled against Cats and Dogs—protecting Yelp's business practice. At the core of this ruling was the Communication Decency Act of 1996 ("Communications Decency Act" 2017).[9] Essentially, the law frees websites to distort the perspective of the world that they confer. Outside a courtroom, we would say the Government now allows websites to lie. More precisely, the Communications Decency Act allows the owner of a website to pick and choose which postings to present, no matter how those selections distort reality.

According to the judge's opinion, any "threat of economic harm [used in business negotiations, associated with the posting of content] is, at best[,] hard bargaining." Platforms like Yelp have the right to "decide what content provided by others to include or exclude." They may "post and arrange [such] content created by others, even if someone may deem such posting or ordering a threat of economic harm." Moreover, Yelp is free to arrange and present content that is true—or false—as long as someone else writes that content. How chilling.

Another example of Government supporting information of a low factual quality involves Facebook:

"Praveen 'likes' Capital One credit cards,"

I've seen this Facebook claim several times. At one level, this advertising claim makes sense. I'm a good friend with Praveen. He knows a *lot* about credit cards, and I respect his opinion. If Praveen liked Capital One, I would listen.

However, Praveen doesn't like Capital One. Praveen works for one of Capital One's competitors and is happy with his current employer.

In fact, Praveen probably never clicked the "like" button on anything associated with Capital One. Facebook's software is designed to identify "shadow likes": things you might like. If one of these things you might like pays Facebook, your friends may see an advertisement suggesting that you actually "like" the advertised thing. There's no requirement that you actually "like" this thing. In fact, you may only find out that Facebook is using you to advertise this thing to your friends if your friends tell you. Facebook won't share (Virani 2015).

In short, our Government supports distorted social discourse. Our Government supports internet-based extortion and it supports false attribution. Such actions warp our view of reality, thereby weakening our social discourse. Support for this type of public news and social discourse interferes with society delivering our needs and wants. This interference is chilling.

Yes, we do receive relevant, factual information from the Internet. However, such factual information tends to burn through the Internet rapidly and then die out. "FDA [Food and Drug Administration] experts offer a unanimous endorsement for pioneering gene therapy for blindness (Carroll 2017)." Such an amazing medical feat! This news did make its way across the internet. However, within a few days, that story was largely gone.

In contrast, what happens with irrelevant or non-factual stories? What happened to the Jade Helm story we discussed a few pages ago about the U.S. government sending the military against its populace? That story is still burning years after it was first told. And what is happening with that perspective on immigration with which we agree? That illustrative perspective is still burning across our personal newsfeeds—and is likely to do so until the day we die.

In short, our public news takes yellow journalism to the next level: it customizes outrage. Social media developers customize the content that's put in front of us to appeal to our personal triggers. Social media shows stories about child abuse to the mother; stories about animal disfigurement to the animal lover; and it is bursting with headlines highlighting political outrage for the grandmother sick to her heart, longing for Walter Cronkite. Social media does not identify the most important stories or provide a coherent context for what we see. And, it leaves us exhausted. Public news is not doing its job in helping society deliver a fair, objective, social discourse.

There are other sources of public information, however. Academics, just like newspaper reporters, in particular, have much to say. What has been the impact of D-Tech on these academics?

ACADEMIC DEBATE

Academics researched much of what is discussed in this book. Much of what they write is insightful. Yet, how constructive is social discourse created by our academics?

Our professors today must fight for our attention just as much as social media and the news must. How do they get attention? The same way your father did. Your father looked up at the sky and said, "I see Elvis!"

No. It wasn't the ghost of Elvis; it was a cloud that looked vaguely like Elvis. However, your father got your attention. Look carefully enough at the clouds and you too will see amazing things: There's a key! And an elephant dancing with a bear. Over there are your Aunt Noemy and her little dog Powder Puff. No. Those items aren't flying through your sky. They are clouds that look somewhat similar to those objects. However, if you have enough random shapes close together, you can at least pretend to see amazing sights.

For at least the past 100 years, we've pressured academics to find the spectacular and amazing things such as the ghost of Elvis. D-Tech gives these academics the ability to rapidly sift through our mountains of data to find any shape desired. So, guess what academics find?

Elvis. Again and again.

Why must our academics fight for attention? Academia is described as creating "a natural selection for bad science (Edwards and Roy, 2007, 52, table 1)." We reward academics for producing more publications. With D-Tech, however, it's particularly straightforward to create "an avalanche of substandard, incremental papers." And we reward researchers for being cited more despite it being straightforward to add meaningless citations with D-Tech.

Figure 3.10: We have as much data as there are clouds in the sky—and, we can imagine amazing stories in each.

We've designed the academic system to spit out attention-grabbing research and it does. The vaccine for measles, mumps, and rubella (MMR) causes developmental disorders in children (Wakefield et al. 1998)! Frightening! But, untrue. The article was later retracted (Editors of The Lancet 2010). Cell phone use causes cancer and low sperm count (Wolchover 2011; Pappas 2012)! Attention grabbing, but false. Most of these surprising discoveries are outrageous because they are *wrong*. Similarly, most claims that "[Artificial Intelligence] machines actually beat doctors" today are also manipulated exaggerations (Oakden-Rayner 2016). But surprise and outrage sells so that's what researchers provide us.

Yes, some research is poor despite being neither surprising nor outrageous. Many academic papers use "poor methods" that lead to "an increase in false discovery rates." They are subject to "reduced quality of peer review." And meanwhile, researchers provide "extended reference lists to inflate citations" and "reviewers request citation of their own work through peer review (Edwards and Roy 2017, 52, table 1)."

As early as 2005, Kurt Kleiner reported, "most scientific papers are probably wrong." More recently, Adam Cifu (2016) of the University of Chicago's School of Medicine reported that, "35%–40% of the findings presented in the prestigious *New England Journal of Medicine* are likely to be reversed within 10 years." The company Amgen failed to replicate 47 out of 53 published cancer studies (Kaiser 2016). Most artificial intelligence research is likely not reproducible (Britz 2017). Scientists find it challenging to reproduce their *own* results because their methods aren't recorded precisely enough—even if their research is not inherently flawed (Lithgow, Driscoll, and Phillips 2017). And more than 60% of published academic journal articles in psychology are either flawed or so narrow that even Ph.D.

psychologists are unable to duplicate the findings (Gilbert et al. 2016).

There's more to this story. We also drive academics to fund their own research. The response? Google, for example, maintains a "little-known program to harness the brain power of university researchers to help sway opinion and public policy (Mullins and Nicas 2017)." In one example of such a program, Google pays academics associated with the discussion of Google's antitrust situation a stipend of $5,000–$400,000. The academics are told that they can use the money "for anything, including opening a doughnut shop." Hmmm. Donut shop? Yummy. However, most professors likely choose to produce academic research. Of course, the professors understand that if they don't support Google's aims then "[Google] will never give [these professors] any more money." Google similarly holds a golden carrot over a number of think tanks (Taplin 2017).

It's not just Google. Coca Cola and Pepsi Co, for example, sell vast volumes of sugary beverages. At the same time, America has an obesity problem "fueled in part by the consumption of sugar (Sifferlin 2016)." Nonetheless, Coca Cola and Pepsi Co fund 63 public health groups, 19 medical organizations, 7 health foundations, 5 government groups, and 2 food supply groups. Included in the Coke- and Pepsi-funded groups are the American Diabetes Association, the National Institute of Health, and the Academy of Nutrition and Dietetics (Aaron and Siegel 2017).

That's a problem. At a minimum, members of such funded organizations must feel qualms about publishing research against their benefactors' interests.

The American Beverage Association responded to such claims on behalf of Coke and Pepsi with, "America's beverage companies are engaged in public health issues because we, too, want a strong, healthy America (Sifferlin 2016)." Engagement is critical. It helps society move forward. However, engagement can also be used to pervert. And, we can imagine all too well what perversion of such a system could look like.

D-Tech may have created the currents that led to Business's role in such academic research perversion. Our public discourse is increasingly twisted in so many ways. Perhaps adding one more twisted story was simply too enticing and straightforward for Business to pass up. The roots of such Business-driven social discourse did exist before D-Tech was invented.

It's also possible this particular, Business-driven perversion of our social discourse is unrelated to D-Tech. It's possible this perversion, with a separate origin, simply pushes in the same direction as the many other D-Tech *side effects*. We'll return to this discussion in Chapter 5.

Not all academic research is low-quality or perverted. Many academics take great pride in their high-quality, frequently spectacular, research. There are many insightful journal articles published. Technology, for example, is progressing because some of the most amazing discoveries are happening right now. However, we are also increasingly overwhelmed by, "Look, son, I see Elvis."

In short, our academic discourse is flawed. Like our public news, our average academic discourse is increasingly distorted and low quality. Much research, itself, is incremental—and even more of it is communicated inappropriately. D-Tech, with its increasing ability to create clouds of information and see Elvis's ghost in those clouds, has played a key role in making this distortion happen.

So, what is the status of our public information? In trouble. Our public information doesn't challenge us in a meaningful way. It panders to us. It shocks us, emotionally exhausts us, and feeds us trivial information that is frequently divorced from reality. It doesn't provide durable, lasting entertainment, nor does it facilitate the meaningful, empowering debate that our society needs so desperately. It doesn't bring us together and help provide for our needs and wants. In short, introduction of D-Tech has weakened our sources of information directly and indirectly.

Recall that social discourse includes experiences as well as information. Let's discuss the types of D-Tech side effects present in our personal experiences.

PERSONAL EXPERIENCE

Our own experiences with the world are the currents that can move our ship to a new location. Rather than anchoring ourselves through social engagement, we drift among specialty friends, family, and work relationships.

How much time did you spend on the Internet and with emails yesterday? Perhaps you had 183 emails and spent 140 minutes on the Internet. That D-Tech time cost you two and a half hours with your family and a day with friends and colleagues. As Marc Dunkelman (2014, 2177) highlights, "For every minute an individual spends on the Internet, the time he or she spends with friends is reduced by seven seconds, and time spent with colleagues by eleven. . . for every email sent or received, an individual lost a minute of time with his or her family." In other words, what happens in our virtual world impacts our physical world.

And what is the impact of these virtual world pressures? Our neighborhood engagement is less in total and more superficial where it does exist. D-Tech has done more to simplify developing relationships with those physically far from us: we can text our friend in Cote d'Ivoire as easily as we can our neighbor across the street. We end up stretching ourselves more thinly and knowing our neighbors less. As sociologist Robert Putnam

(2000, 138) describes, we have "less integration of American adults into the [local] social structure."

Politics, for example, no longer engages as many members of society as it did before D-Tech. "From the earliest opinion polls in the 1940s to the mid-1970s, younger people were at least as well informed as their elders were, but that is no longer the case (Putnam 2000, 75)." We're also not voting: "Participation in presidential elections has declined by roughly a quarter over the last thirty-six years. Turnout in off-year and local elections is down by roughly this same amount (Putnam 2000, 61).[10] Perhaps there is a silver lining in that uninformed people aren't voting. However, our neighborhoods would be stronger if we were both informed and voting.

We are also less socially engaged in general. More and more of our social clubs are like VFW Post 2378 (Putnam 2000, 11):

> VFW Post 2378 in Berwyn, Illinois, a blue-collar suburb of Chicago, was long a bustling "home away from home" for local veterans and a kind of working-class country club for the neighborhood, hosting wedding receptions and class reunions. By 1999, however, membership had so dwindled that it was a struggle just to pay taxes on the yellow brick post hall. Although numerous veterans of Vietnam and the post-Vietnam military lived in the area, Tom Kissell, national membership director for the VFW, observed, "Kids today just aren't joiners."

We may spend an hour on Facebook with friends far and wide. However, we're not socializing with our neighbors. "In the mid to late 1970s, according to the DDB Needham Life Style archive, the average American entertained friends at home about fourteen to fifteen times a year. By the late 1990s that figure had fallen to eight times per year, a decline of 45 percent (Putnam 2000, 254)."[11]

Where are we engaging? We're engaging with our specialty friends and in our specialty business arrangements. We're more selective about where we do engage within our physically-nearby neighborhood. Increasingly, we are forming specialty families by mating with ourselves: Americans were four times more likely to marry someone with their same level of education in the early 2000s, up from three times four decades earlier (Mare 1991, 20, 23–24). A study in the early 1990s found that the odds that a high-school graduate might marry a college graduate fell by a quarter between 1940 and the late 1980s (Mare 1991; Schwartz and Mare 2005, 4).

D-Tech applications like Meetup attempt to bridge the physical and virtual divide. Users of Meetup can find gatherings of people with like-minded interests. And, Meetup does forge new relationships (Sander 2005). However, like many of our current relationships, such ties tend to be shallow. For the most part, our relationships aren't the sorts of ties that span more than a single common interest (Dunkelman 2014, 2128).

What relationships are increasingly missing from our bevy of specialty friends? Those close friends (both physically and with a deep connection) who understand us and help us through life. In 2004, nearly a quarter of Americans reported that they have no one they can confide in at all; more than half (53%) said they could confide only in a family member. Twenty years earlier, that number was only 36% (McPherson, Smith-Lovin, and Brashears 2006, 359, table 2).

Our specialty business arrangements are those with the unique skills we need—right now. Once again, D-Tech helps us find "the precise right answer," and we reduce our use of the generalist. Growing numbers of firms rent consultants who are deeply specialized in their field. The number of subcontractors has exploded as people want to hire experts rather than jacks-of-all-trades (Florida 2002, 53–55): in my virtual Rolodex, I find French real estate tax accountants, Belgian water pollution cleanup specialists, and experts in training plumbers about client communication. Academia is trending the same direction: generalists are increasingly rare, and scholarly specialization has become the new norm (Dunkelman 2014; Walt 2009).

We don't even need to interact with our specialty business contacts as though they are real, live people. To quote Marc Dunkelman (2014, 1943),

> A white Klansman from Kentucky may think a woman's place is standing over a stove. Nevertheless, he might unknowingly trade his rookie Stan Musial baseball card to a black woman from Oakland who owns a small business. [Even when these] relationships pierce the prejudices that once balkanized American society, [they do not address] the underlying tension.

We interact with many such business contacts through low-density, digital communications. The precise business meaning of our email or text is clear, but no emotion shows through. The result would be largely the same if our counterparty was a machine rather than a human.

We have used D-Tech to search out the person with the deepest expertise in relation to our specific problem we could find, however. Thus, to participate in such impersonal Business relationships more of us are getting advanced degrees. This huge rise in professionals is, in turn, growing many professional organizations, as illustrated in Figure 3.11: membership in the American Medical Association (AMA), for example, rose from 126,042 in 1945 to 201,955 in 1965 and then to a record 296,637 in 1995. However, a falling percentage of professionals are actually joining that professional organization (Putnam 2000, 209). Despite surging growth in the number of doctors, for example, membership in the AMA was down to 217,490 in 2011 (Robeznieks 2013). Our connection with even our fellow professionals is weakening.

Yes, we are connecting for common self-interest. However, the "proliferating new organizations are professionally staffed advocacy organizations, not member-centered, locally based associations." And, many are, "headquartered within ten blocks of the intersection of 14th and K Streets in Washington: the Children's Defense Fund, Common Cause, the National Organization for Women, . . . the Wilderness Society, and Zero Population Growth." In other words, these new organizations are taking advantage of the growing opportunity to advocate Government, discussed in Chapter 4. These are not primarily relationship-building organizations. As Putnam described, "the 'new associationism' is almost entirely a denizen of the Washington hothouse (Putnam 2000, 115)."

Our broader network may help us find work (Mills 2013). That's positive. Unfortunately, D-Tech is also causing some *side effects* on our labor market that are negative. No, for the time-being, D-Tech hasn't replaced us all, although we can't rule that out for the future. The *side effect* is that millions of American youth and prime-aged males are unemployed today, a topic to which we'll return in Chapter 5.

In short, D-Tech has done more to help us develop relationships with those physically distant from us and to develop niche knowledge. Our physical engagement with many of the living, breathing, feeling human beings close to us is shrinking.[12] Instead, we are developing specialty and often virtual relationships with friends and business partners. Rather than developing robust, deep relationships that broaden our perspective, we're finding everything we need in the moment—without broadening our perspective.

Participation in Professional Organizations

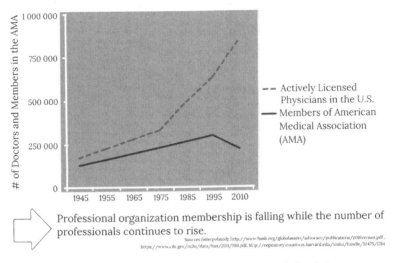

Professional organization membership is falling while the number of professionals continues to rise.

Sources (interpolated): http://www.fsmb.org/globalassets/advocacy/publications/2016census.pdf , https://www.cdc.gov/nchs/data/hus/2011/109.pdf, http://repository.countway.harvard.edu/xmlui/handle/10473/1784

Figure 3.11: While growth in the number of physicians in the U.S. has continued to grow, membership in the physician's professional organization, the AMA, has declined. Other professional groups and social clubs across the U.S. exhibit similarly declining membership.

SUMMARY

In summary, Culture has responded poorly to D-Tech. Rather than rising to the challenge that D-Tech has created, each aspect of our social discourse is increasingly compromised. First, our popular information is distorted. To maintain our interest in two-minute blocks of time, our social media is repeatedly throwing hooks at us rather than informing us. It's insulating, emotionally exhausting, and trivially combative. Social media also supports our further warping of information through the ability to freely comment and repost. D-Tech also makes it easier to maintain a broad range of contacts—but more challenging to deepen our friendships.

Business and Government are reinforcing these destructive Cultural responses. Our popular news industry is insufficiently motivated to provide quality content, while our Government is hosting close-hold news embargoes and supporting internet-based extortion and false attribution.

Second, too many academics are pointing at clouds and telling us they found Elvis. Yes, there are constant, groundbreaking insights. However, much academic research is neither novel nor accurate.

No, D-Tech did not destroy a previously pristine social discourse. For example, we've been waging a 65-year war against saturated fats based on bad science (Teicholz 2014), when better science has existed for much of that time.

However, the negative, destructive aspects have accelerated with D-Tech. Our social discourse is filtering information less effectively, and higher-quality discussion and debate are less common. There is an increasing gap between what we need from our social discourse and what it delivers.

Our social discourse exists to anchor our relationships and our cultural norms. But, it's not. It is not providing a coherent view of our world. It's not providing the quality information with sufficient background needed to analyze our world well. And, it's not guiding constructive, relationship development.

What is the impact of these changes in our social discourse? That is the topic of the next section.

WHAT IS THE IMPACT OF CULTURAL ADAPTATION?

Our Culture is failing to support society.

That's a strong statement. It's also a fair statement.

How do we judge Culture? We judge Culture as a subsystem of society. The purpose of Society is to deliver our needs and wants. Culture's purpose within this broader system is to help us respond constructively to our world. Its cultural norms help us think less and act more: if those actions help society deliver our needs and wants, then they are good, cultural norms. Culture's relationships nourish us directly and provide resources for us to solve problems. If these relationships are nourishing and helping us solve our problems, then these are good. Many of our cultural norms and relationships are not fulfilling their purpose, however.

As we discussed in the last section, the *side effects* of D-Tech have damaged our social discourse. Without constructive information, experiences, and standards we're destined to find damaged cultural norms and broken relationships. No computer is reaching out its hand to trip us as we run. No digital phone is giving us an electric shock if we reach a hand out to help a neighbor. Nonetheless, the breakdowns in our Culture are *side effects* of our applying D-Tech:

- D-Tech's capabilities are impressive: unparalleled access to information, unrivaled capacity to analyze data, and exceptional ability to influence decisions.
- As we apply these capabilities, we are changing our environment: creating a virtual world stuffed with masses of data vying for our attention, while also increasing the complexity of our physical world.
- Our Culture is breaking down as a result:
  - Many across society are creating sticky, attention-grabbing content. Meaningful discourse is suffering.

- We have access to lower-quality news, less capable of cutting through our world's complexity to provide us insight and quality discourse.
- We are less integrated into the physical society where so much of our life occurs.

As we discuss in this section, we are suffering as a result. Our Culture, today, is failing us both as individuals and as a group.

MANAGING OUR INDIVIDUAL NEEDS AND WANTS

Our Culture is failing individuals across society. As we walk around feeling dazed by the world's complexity, we feel that something is wrong. Our Culture seems to be betraying us: it isn't helping us satisfy our needs and wants. Over the coming pages, we confirm the failure of our Culture to satisfy our needs at every level of Maslow's Hierarchy of Needs (Figure 3.4).

The base of Maslow's Hierarchy is delivery of our physiological needs. We rely on cultural norms to help teach our children how to address such needs. Where are we in teaching our children how to approach the world's physiological problems? Have you seen the Home Depot (n.d.) video about how to use a tape measure? Home Depot spent millions of dollars producing basic videos on how to use tape measures, how to mop floors, and how to hit nails with hammers (Bryon 2017; Durden 2017a). Yes. Really. Home Depot executives were concerned such topics would be too condescending. As it turns out, those concerns were unfounded; the videos are a hit.

Just how many of these basics is our Culture missing? One in five adults is unable to boil an egg or change a light bulb. Two in five adults cannot unclog a sink or change a diaper (Aviva 2017, 14). And, no, we still don't have robots that can do those tasks.

We might ask, however, whether egg boiling is simply too basic for today's society. Perhaps our Culture has simply shifted its focus higher up Maslow's Hierarchy of Needs. No, we are struggling at Maslow's higher levels, as well.

We want a safe society. We would love for our Culture to support safety; that's why we sing lullabies with words like, "I would wrap you in angels, just to keep you safe (Picard 2006)"!

Our cultural norms are *not* keeping us safer, despite a "Great Crime Decline" across the U.S.. The "Great Crime Decline" is a decrease since 1990 in our violent and non-violent physical crime rates. Some analysts have suggested strengthening cultural norms explain this decline. Ford (2016), for example, explains how no other change in our physical world could explain such a crime decline: not a change in the number of police, nor falling alcoholism, nor improving household anti-theft technology. If it's not those other aspects of our physical world, then it must be strengthening cultural norms and relationships. Right?

What a wonderful story—but it's wrong. If you were operating an organized crime syndicate today, would you focus on the physical world where we increasingly have cameras to catch you? Or, would you focus on the virtual world? A gentleman whom others have suggested is involved in organized crime called me the other day. He didn't threaten me or tempt me to take drugs. His request, "Tell me about the blockchain." The blockchain is a powerful, new Digital Technology toolbox.

Physical crimes began to fall in 1990 as the internet exploded. The internet has become a superfood for virtual crime. Hackers spy on us, increasingly through an expanding array of our life: smart phones, apps, laptops, baby monitors, TVs, cars, and more (Haynes 2017). Between 2004 and 2016, 28% of students were cyberbullied, and 16% of students cyberbullied someone else (Patchin 2016). In 2014, cyber hacking exposed 783 million records in the United States (Akpan 2015), more than two records for every person in the country! In 2015, 85% of all organizations were victims of phishing attacks (Crowe 2016). Twenty million online images a week are preliminarily identified as child pornography featuring children under 12 years old (Thorn, n.d.). Every year, millions of scams are perpetrated through email, websites, and social networking (ACCC, n.d.).

Amazingly powerful cultural norms and relationships are not helping society reduce crime. It is, at best, a sweet sorrow that our cultural norms may be helping us respond to this less-safe world. Americans think society is plagued by increasingly high, violent crime rates (Eisen and Roeder 2015). That belief may be perfectly aligned with our current, D-Tech reality: cyber-crimes are increasingly violent even though they are not physical. No, our Culture is not helping improve our safety.

How about our social connections? That is the next level of Maslow's Pyramid. "You can now communicate with anyone, anywhere, at anytime." That promise is what sends us off to purchase cell phones for our children and loved ones. The promise is that the cell phone will strengthen our bonds and make it simpler to coordinate school pickup. And boy, have we been communicating! One 2015 source indicates that every minute of the day, for example, "email users send 204,000,000 messages," and "Skype users connect for 23,300 hours (Kimmorley 2015)."

Unfortunately, corresponding, cultural norms would be needed for such spectacular, communication technology to truly bring us together. We would need cultural norms that supported our using D-Tech when its use would bring us together—and not when its use would draw us apart.

We think that our online status is our real status—until no one in the real world appears to care. We maintain a bevy of superficial relationships. However, as highlighted earlier, we're not maintaining close friendships or relationships with our neighbors. An AARP study found that 35% of those 45 and over reported being "chronically lonely," versus 20% in that state only 10 years earlier (Edmondson 2010).

Figure 3.12: Almost 40% of adults reportedly don't know how to boil an egg. And, no, we still don't have robots to do that task.

Above social connection, Maslow placed self-esteem. Is our Culture helping us strengthen our self-esteem? No. Our self-esteem is falling. The suicide rate among 15–24-year-olds has tripled since the 1970s (AFSP, n.d.). The suicide rate for middle-aged men increased 43% between 1999 and 2014 while the rate for middle-aged women increased 63% (Curtin, Warner, and Hedegaard 2016). The suicide rate for the elderly has fallen. However, we just discussed the chronic loneliness of the elderly. It seems, then, that the elderly are simply not pulling the trigger despite their chronic loneliness.

Finally, our Culture is not supporting Maslow's highest level: our self-actualization. Our Culture has become more twisted. Talent used to bring about fame. Robin Williams, for example, earned his fame for over 30 years by entertaining audiences with exceptional comedic talent. Unfortunately, social media has provided more powerful methods for developing fame than mere talent: fame itself begets fame. Simply having "likes" and followers begets more "likes" and followers. The Kardashians. Need we say more?

More recently, our Culture has taken another twisted step: Fame is increasingly earned by creating outrage. Our world, and particularly social media, is in a fight for attention. Hate and outrage attract attention. Thus, it is hate and outrage that begets fame. The more hateful and the more outrageous someone acts, the more likely that person is to be rewarded with fame. I have no interest in providing additional coverage to those who have become famous for delivering outrage and will provide no examples here.

It is possible to take this perspective on self-actualization too far. There certainly remain people who become self-actualized for the talent and benefits they bring society. Richard Thaler won the Nobel Prize in economics in 2017, not one of the

Kardashians. Pop culture is where the twisted trend is be-coming more common. Perhaps pop culture has never been particularly healthy. However, more fringes of this culture are especially unhealthy today. Our Culture is not increasingly sup-porting self-actualization.

In addition to the negatives, there are also positive develop-ments in our Culture. Some aspects of Culture have progressed. We have made strides, for example, in integrating people with physical disabilities into mainstream society. Throughout much of the 20th century, these people were, "pitied, ridiculed, rejected, and feared, or [treated] as objects of fascination." Increasingly, today, there is a "growing assumption that persons with disabili-ties should be full and equal participants in all aspects of society (Neuhaus, Smith, and Burgdorf 2014)."

Other areas have shown progress as well. More people are engaged in cross-cultural relationships. In 1964, only 18% of whites claimed to have a black friend. By 1998, 86% of white people said they had a black friend, and 87% of blacks said they had a white friend (Thernstrom and Thernstrom 1998).

These triumphs, however, are localized. The gap is growing between what we need our Culture to do directly for us and what it is doing. It is not ensuring we can teach our children the basics of physiological life. We are not feeling safer. We have weaker social connections. A large portion of the population suffers from loneliness and low self-esteem. Our Culture is not helping us leverage Digital Technology to deliver more of our needs and wants.

SUPPORTING OUR TEAMWORK

Our Culture is also failing our collaboration. Our neighborhoods, towns, and cities are all slower and less effective at problem solving. Our ability to band together to solve common and complex problems is critical to our success as a society; our success depends on our being a collaborative species. Nonetheless, that ability is weakening.

Where were we then, and where are we now? Tales of American Revolution heroes, for example, leave us awed. Paul Revere rode his stallion throughout New England:

> *One if by land, and two if by sea; And I on the opposite shore will be, Ready to ride and spread the alarm Through every Middlesex village and farm, For the country-folk to be up and to arm (Henry Wadsworth Longfellow. 1860. "Paul Revere's Ride").*

However, behind such inspiring individuals "stood the networks of civic engagement in the Middlesex villages (Putnam 2000, 43)." Perhaps it was this network and the Culture.

Today, our Culture hinders our ability to solve collective problems. On the one hand, the complexity of our Culture slows the problem-solving process. Culture, itself, for example, constantly needs rejuvenation. Yuval Harari (2015, 163–64) explains, "Cultures... undergo transitions due to their own internal dynamics. Even a completely isolated culture existing in an ecologically stable environment cannot avoid change. Unlike the laws of physics, which are free of inconsistencies, every [society] is packed with internal contradictions. Cultures are constantly trying to reconcile these contradictions through social discourse, and this process fuels change."

This Culture rejuvenation process has become more challenging with D-Tech. Consider women's rights. The trend through the early part of the D-Tech era was clear: society was increasingly prioritizing women's rights over the nuclear family:

> *1960: Sonja dreamed of saying to Jim, "I'm frustrated. We have three kids and I spend my day cleaning and washing." However, it would be tough for Sonja to get a job, and even tougher to get a divorce. Her dream continues, "This isn't fair." Sonja might be passive aggressive. She might hate Jim. However, Sonja knows that society has prioritized the nuclear family over her rights. She plays her role (reasonably) quietly.*
>
> *By 1975: "I'm leaving you, Jim. We have only one child. I have a job and I can divorce you no matter what you say. Ha!" and Sonja is gone.*

In the 1960s, a woman was expected to follow one path: "to marry in her early 20s, start a family quickly, and devote her life to home-making (Tavaana, n.d.)."

Between 1960 and 1975, women's rights rapidly developed: Sonja now had choices. Technology provided birth control in 1960 giving women power over family planning (Thompson 2013). In 1970, California's Government began the trend of supporting female empowerment by implementing no-fault divorce (Jaworski, n.d.). Now women could end a marriage without their husband's agreement, and with neither party taking the blame for the end of their combined family. Then, in 1973, Government (the Supreme Court) legalized abortion, further strengthening women's control over family planning. Meanwhile, as D-Tech expanded in the workplace and the memory of Rosie the Riveter became an endearing myth, more women performed a wider variety of factory jobs, as well as other jobs.

With D-Tech, this process of shifting Cultural priorities essentially came to a grinding halt. We got stuck with inconsistent expectations—and, even more importantly, we got stuck debating tangential items while some of the core, most impactful items remain unattended. Women have the rights and many of the capabilities to leave their husbands. Yet, too frequently, women have to choose between themselves and their children. Divorced parents tend to obtain less education and have lower incomes. Perhaps as a result, children raised by single parents are more likely to fare worse on a number of dimensions: school achievement, social and emotional development, health, and labor market success. These children are at greater risk of parental abuse and neglect (especially from live-in boyfriends who are not their biological fathers), are more likely to become teen parents, and are less likely to graduate from high school or college (Sawhill 2014).

As a society, we began the process of prioritizing women's rights over the nuclear family decades ago. However, we're still miles from developing a Culture consistent with such a priority. If women's rights are to be a priority, single parents consistently need support to raise their children well. Certainly, many single parents do this alone, but it is highly challenging. Society hasn't reprioritized the nuclear family over women's rights. However, we are stuck between the two, arguing heatedly about aspects frequently tangential to what is critical—and our children are suffering as a result.

We spend lots of energy developing our Culture. However, this energy is not helping us make progress on the fundamental questions such as whether women or our nuclear family should take priority. There is massive friction and debate about women's rights, to be sure. Nonetheless, many of the basic, fundamental actions that could have the most beneficial impact on women, men, and their children remain undone. It is in these areas where society's greatest focus should be: on collaborating to ensure such high-value, high-impact needs are covered. Too frequently, instead, we are simply sitting inside our cultural bubbles, supporting arguments that allow us to maintain consistency with our side, while contradicting the views of our neighbors.

Let's move from the rights of individuals and families to examining our ability to collaborate. Neighborhoods mean teamwork! That seems like a nice saying. However, it would be an empty expression today. We've damaged the infrastructure that makes many neighborhoods, cities, and towns work well. "Every movement in American history has cultivated and mobilized people who might not otherwise have gotten engaged in the political process (Dunkelman 2014, 1932)." American social movements have all been about civic connection and relationship. Today, where are we with our neighbors?

> *"He waved," Ming said. "The next door neighbor smiled at me and waved."*
>
> *"You're kidding. We've lived here for six years, and he's never said a word to us," I replied. "Are you sure that was really him? I hope he's not sick."*

Yes. That was a real conversation. "Today, being 'neighborly' means leaving those around you in peace . . . The sense of warmth once suggested by the term . . . has been replaced by a kind of detachment (Dunkelman 2014, 1945)."

Without knowing our neighbors, we cannot understand them—and we certainly cannot collaborate easily with them. We communicate less and less with anyone living physically close to us. The time we used to spend chatting over the back fence is now spread across our 600 virtual friends on Facebook. Without communicating with our neighbors, it's difficult to work together on complex tasks.

Technology has allowed us to work together with our neighbors on tactical activities. For example, the application "NextDoor" allows neighborhood users to share information about prowlers and recommendations for local plumbers. Where we need only superficial relationships with each other, such applications can be tremendous (e.g., "Watch out for the green, 1970s pickup truck because the driver is stealing lawn gnomes.") However, such applications do not consistently help collaborate on more complex issues such as how we need to rezone our town. Such applications don't help us develop a coherent view of our neighborhoods or our neighbors (Stross 2012).

We'll return in Chapter 5, for example, to discuss complex neighborhood problems that remain unsolved. Such neighborhood challenges may be costing us 40% of our economic production.

So, what is the impact of Culture's adaptation to D-Tech? Our Culture is not playing its role in helping us. As individuals, we are failing to obtain what we need at every level of Maslow's Hierarchy. Culture is hindering our ability to solve the most complex neighborhood problems. And, D-Tech is slowing the process of collaborating on many of the most critical issues to our society.

SUMMARY: IMPACT OF DIGITAL TECHNOLOGY ON OUR CULTURE

Digital Technology has transformed Culture. *Directly*, Culture clearly benefits from D-Tech. We can now consistently contact our loved ones, for example, and it's easy to maintain contact with people from our past.

However, D-Tech also has *side* (or emergent) *effects* that appear in our Culture. D-Tech pervades our culture as its sea of information and incredibly fine filters set the framework within which we interact. We may have lost sight of our next-door neighbor—yet we know when someone we haven't seen in 30 years finds a particularly big shrimp on her dinner plate.

Meanwhile, D-Tech, many times seemingly without human intervention, sets the bias and framework from which we view the world. D-Tech biases us. It provides unmitigated potential to create expansive amounts of imprecise information: the result is a vast data smog and destructive, personalized fight for our attention. As a result, too frequently, D-Tech leaves us depressed. We are tied to our social media sites as they successfully distract us from the rest of our lives. Meanwhile, D-Tech's ability to record customized data has allowed us to create complex options and recognize more special interests vying to impact our world.

We have not actively re-designed our Culture to manage these new challenges as successful leaders of the Agricultural Revolution did. We have not benefited from a Hammurabi, for example, who emphasized the newly relevant, personal responsibility for public works.

Rather than rising to these D-Tech challenges, each aspect of our social discourse is increasingly compromised. Our popular information must maintain its hooks for our full two minutes of interaction; it frequently earns our mistrust. What we receive is insulating, emotionally exhausting, and trivially combative, AND, it supports our further warping of the information we find through D-Tech.

Now it's easier to maintain a broad range of contacts—but more challenging to maintain deep friendships. Business and Government are reinforcing such destructive social discourse. Our popular news industry is insufficiently motivated to provide quality content, while our government is supporting internet-based extortion and false attribution. Yes, we have truly groundbreaking research and some aspects of our culture are constructively developing. However, many academics are chasing clouds.

Overall, there is an increasing gap between what we need from our social discourse and what it delivers. Our social discourse is not providing a clear and coherent view of our world. Nor is it fostering constructive, relationship development. Instead, it is filtering information less effectively; as a result, higher-quality discussion and debate are less common.

With our Culture faltering, our society is less effective. As individuals, we are failing to obtain many of our needs at every level of Maslow's Hierarchy. We have the perfect contact for every activity, but we're lacking the friendships that help us find our way through life. Too many people are lonely, addicted to drugs, or resorting to suicide. Too few of us are aware of just how staggering these problems are.

As neighbors, we're missing the relationships needed to solve problems. We're increasingly slow at aligning contradictions within our society. We can solve the little problems, but larger, higher-value personal and neighborhood issues remain unresolved.

In many respects, we are unlike the successful Agricultural Revolution societies. We are not progressively evolving our Cultural institutions in the face of the increasing challenges we face. We are sowing the seeds of potential in our new virtual world—without updating and upgrading the Cultural structures so they will nurture these efforts. Our Culture is drowning in potential.

In our next two chapters, we will continue to explore the impact of D-Tech on our society. Are the benefits we obtain from Government progressing as that sub-system actively helps us leverage D-Tech? Or, is this system and outmoded approaches to collaboration being overwhelmed by D-Tech just as in our Culture? We need to know.

CHAPTER 4

THE IMPACT ON AMERICAN GOVERNMENT:

Lost and Confused in the Digital Technology Age

In this chapter, we look at the impact of Digital Technology (D-Tech) on Government. The perspective in this chapter is informed by my experience as an invited researcher to the Japanese Ministry of Trade, Economy, and Industry. The bulk of people I met were warm, inviting, and dedicated to serving their nation. I expect the same is true of the U.S. government as well. There are certainly individuals who undermine policy initiatives. However, we will learn more about Government's failures by studying the overall system than we will by studying such rogues.

D-Tech has allowed us to develop tax databases, analyze education results, and offer impressive websites to inform citizens. Such innovation is wonderful.

However, our government is failing along too many critical dimensions, as well. In this chapter, we investigate the relationship between D-Tech-driven complexity and our governmental failures. Government is constantly upgrading sophisticated websites and databases. Yet, the outcomes are not what society needs. These challenges are illustrated by government's regulations. Government is not using technology solely for regulations that protect us from our most dangerous, animalistic instincts. Instead, regulations are increasingly sucked into D-Tech-driven complexity with the average professional facing:

> *Three felonies a day: The U.S. Federal Government is enforcing more than 250 million restrictions on what we do.[1] Many of these restrictions are so complex and vaguely worded that the average professional, without realizing it, commits several federal crimes every day (Silverglate 2011, xxv). Meanwhile, almost three-quarters (74%) of Americans report being afraid or very afraid of government corruption (Chapman University 2017).*

As we will discuss over the course of the chapter, government is decreasingly effective in its role: setting the stage for successes of our culture and business/economy. Like our Culture, our Government is suffering from *side effects* (i.e., *emergent impacts*) from our application of D-Tech.

As we will discuss in this chapter, over the past 50 years, the impact of D-Tech on Government has exploded. No longer is D-Tech providing a gentle current of support for Government efforts. Government today is grappling with our increasingly complex world—that we have only been able to create thanks to D-Tech's tools. As we are able to change our world more fundamentally and broadly through technology, more areas certainly require Government constraints and requirements. Yet, the current massive complexity of regulation, public works, and Government efforts is stifling, rather than supporting.

Our Government's experience with D-Tech contrasts with the experience great societies of the past have had with their technology. Thousands of years ago, for example, great leaders professionalized Government to respond to Agricultural

Technology. As Agricultural Technology was increasingly applied, nomadic societies became settled. As Agricultural Revolution settlements grew, Government in successful societies developed systems to protect themselves from invaders and miscreants. Trade was supported by legal systems that enforced agreements, and thriving economies required the shared public works of dams, levees, and roads.

Today, we are not constructing a Government able to support society through our current challenges. Government may be protecting us from invasion and some of society's most destructive individuals. However, our Government has lost its role as a leader of society. We are embroiled in a sea of confusing regulation and legal precedent. Our public works departments are expensive to run and maintain, yet in many ways, mediocre. And, as a nation, we are borrowing as if we are in the midst of a war—in one of the longest peacetime eras in our history.

Certainly, Government benefits *directly* from D-Tech. A computer is assigned to every bureaucrat on their first day at the job.

Nonetheless, D-Tech's *indirect impacts* (also called *side effects* or *emergent impacts*) on Government are even more powerful. As discussed in Chapter 2, D-Tech harms our ability to think clearly. And, when we can't think clearly, we govern our actions poorly. In the movie theater, we'd have a pretty good idea what follows the line:

"Richard! Don't chug those martinis before heading to the party."

In the movie, Richard would chug the martinis. He'd go to the party drunk and his thinking would be fuzzy. And, unable to think clearly, he'd govern his actions poorly:

- He'd talk too much, but have no clear plan.
- He'd be too easily influenced to do something he should not.

- He'd get lost on the way to his destination, perhaps causing injury.

We've seen what happens when a drunk Richard leads the U.S. Government: President Richard Nixon was supposedly drunk on several key occasions during his presidency. On those occasions, strong individuals had to stand in for the President.

Reportedly, Richard Nixon ordered plans for a tactical nuclear strike against North Korea in 1969, for example. North Korea had shot down an American spy plane, and Richard was angry—and drunk. U.S. air force pilot Bruce Charles was "put on alert to drop a 330-kiloton nuke on a North Korean airstrip (Stilwell 2017)." Then National Security Advisor, Henry Kissinger convinced the Joint Chiefs of Staff to stand down on that order until the President could wake up sober in the morning.

Similarly, during the 1973 Middle East War, then Secretary of State Henry Kissinger received a message from the Secretary General of the Soviet Union's Central Committee, Leonid Brezhnev: "I will say it straight. If you find it impossible to act jointly with us...[we will] urgently consider taking appropriate steps unilaterally." Unfortunately, President Nixon "could not be awakened. Nixon obviously had had too much to drink (Stilwell 2017)."

"American intelligence confirmed suspicious Soviet air and naval movements. . . At US air bases, B-52s loaded with nuclear weapons lined up nose to tail. In missile silos, launch commanders buckled themselves into their chairs. . . A stern reply to Brezhnev's letter. . . went out at dawn. . . sent in Nixon's name (Summers 2000)." And, all of these activities occurred while Richard Nixon, the leader of the free world, slept off the booze.

Figure 4.1: While drunk, then-President Nixon once put a U.S. bomber on alert to drop a nuclear bomb on North Korea. When Government leaders' thinking is fuzzy, the potential for problems grows.

In many ways, our Government is like a drunken Richard whose thinking is fuzzy. The impact of our Government suffers from challenges Gathering Information, Understanding Context, and Analyzing Options. As a result, our Government is working poorly. Those with clarity and the strength to do so are stepping in and making many of the decisions that are coming out under the Government's name. Unfortunately, few of those with the strength to influence Government decisions today are exhibiting the decision-making quality that Henry Kissinger did for Richard Nixon.

Society is an adaptive system, constantly adjusting in the face of external constraints and opportunities. Over the course of this chapter, we focus on the second-order *side-effects* of D-Tech felt by Government. A systems scientist would call these impacts emergent: arising as a result of D-Tech, yet not uniquely traceable to that origin.[2]

This discussion covers the grey-highlighted portion of Figure 4.2: as Society evolves, Government responds to changes in the environment in which we live and work (Arrow 1). In particular, as we discussed in Chapter 2, D-Tech is creating an environment in which it is more challenging to think clearly. How our Government responds also depends on A) changes in the Businesses that lobby it and B) shifts in the Culture of those who vote it into power (Arrow 2).

So, what is the most obvious, indirect impact of Digital Technology on our Government? D-Tech has made our lives more complex, leading to a more complex Government. Consider the activities of the U.S. Department of Agriculture (USDA). According to Google, the USDA has about 106,000 employees today. That's larger than the average Fortune 500 company.[3]

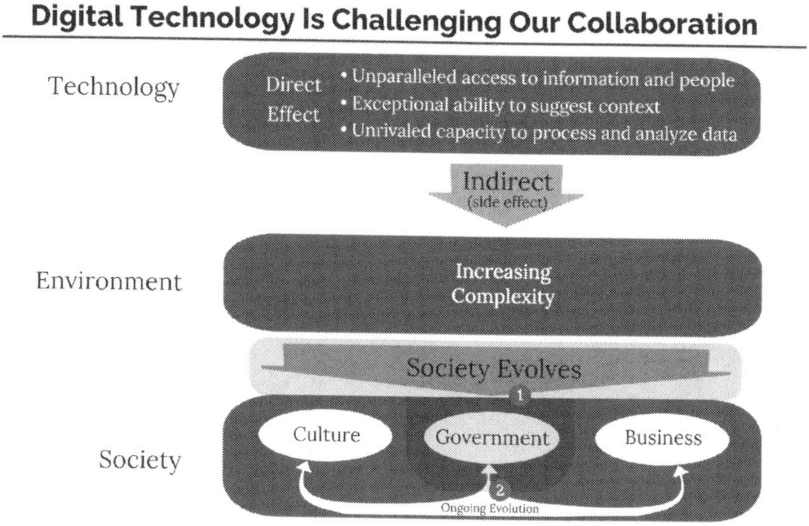

Digital Technology Is Challenging Our Collaboration

Technology — Direct Effect
- Unparalleled access to information and people
- Exceptional ability to suggest context
- Unrivaled capacity to process and analyze data

Indirect (side effect)

Environment — Increasing Complexity

Society Evolves
1

Society — Culture · Government · Business

2
Ongoing Evolution

Figure 4.2: American government has evolved
as a result of D-Tech-induced complexity,
i.e., as a result of D-Tech side effects.

The USDA investigates dog fighting (and arrested Michael Vick [Animal Legal Defense Fund, n.d.]). It monitors the plague, rabies, and feral swine disease (Animal and Plant Health Inspection Service 2017a). It manages vultures; in addition to managing the salmonella and other diseases the birds carry, these actions also help reduce the amount of roofing caulk the birds eat (Animal and Plant Health Inspection Service 2017b). The USDA shoots fireworks at Canada geese outside airports to ensure they don't clog airplane engines (Lewis 2017). And, of course, it monitors, researches, and oversees subsidies to American farmers.

What does this complexity look like from inside the USDA? Some USDA employees play drinking games using their employer as the target (Lewis 2017). One will ask the other, "Does the

USDA do _____?" Responders drink when they get the answer wrong. They get soused.

Despite such a broad set of activities, the USDA's purview "Agriculture" only represents 1% of the U.S. economy (Economic Research Service 2017).[4] The U.S. federal Government has 249 agencies in addition to the USDA, and 2.8 million federal Government employees in total. If we put all those employees together, we'd have a city the size of San Diego, California. (And that's not counting the 0.5 million employees of state and local government.)

The U.S. Government has become mind-numbingly complex. I make no judgments about whether Government should be large or small. What I care about is the benefits society obtains (or fails to obtain) through such a complex Government. What *side-effects* are we experiencing?

We answer this question here, dividing this chapter into three sections. First, why is there a relationship between D-Tech and how well our Government delivers benefits to society? Here, we expand on the description of our increasingly complex environment that we discussed in Chapter 2, focusing on those aspects relevant to Government. Second, how has our federal Government adapted to this complexity? In short: poorly. It's acting like a drunk "Richard" trying to prepare a complex scenario for others. Third, what is the result of federal Government adaptation? My answer: have you ever seen Richard throw up?

I judge Government's success on how it sets the stage for society by:

- Effectively setting the limits of acceptable behavior through regulation.
- Providing infrastructure (public works).
- Managing its finances, balancing the needs of today and the future.[5]

WHY IS D-TECH STRESSING OUR GOVERNMENT'S FUNCTIONING?

D-Tech changed our environment. Governing today requires gathering global, complex information. Putting that information into context despite increasing potential for voter manipulation. And, predicting how particular actions are likely to impact society—despite the complexity of doing so.

GOVERNMENT GATHERING OF INFORMATION

Consider Government's challenges with Gathering Information in the age of Digital Technology. It faces a swarming sea of data created by our 2 billion computers. This information is coming from around the world, essentially without borders. As a result, the volume of nuanced information our Government must understand is intimidating.

Have you ever heard of the molecules called 3-MCPD esters? The full name of these molecules is 3-monochloropropane diol esters. These molecules are... Let's be real: I have no idea what 3-MCPD esters are.

I used to be able to wave my arms and cluck like I knew what 3-MCPD esters were. The biochemists explained these molecules to me; my most important business project for several months focused on them. However, the language biochemists speak is foreign to me.

I did learn some things about 3-MCPD esters, however. These molecules are a concern: they showed up in infant formula tests around 2010, and they harm infant brain development. My children drank infant formula free of 3-MCPD esters—and I hope everyone else's babies did/do, as well.

Regulators need to make rules in such situations. However, communication about these rules can be frustrating, and miscommunication can lead to serious consequences. Consider illustrative comments that would come from knowledgeable biochemists to regulators.

To make the appropriate decisions in this example, the Government needs to be fluent in the languages of both biochemistry and the infant formula supply chain.

We need a Government able to understand all of the languages of today's world. And this understanding must come before Government can even begin to consider what actions to take and whether to remain independent of the special interests that are fluent in these languages.

We now turn to the second challenge.

GOVERNMENT UNDERSTANDING OF HOW OUR WORLD OPERATES

To fully understand the information we gather, we need to put that information into context. In the case of voting, we might disagree with a particular candidate's views on international trade, for example. Yet, that disagreement should be placed into a broader context—how much we agree or disagree with the candidate's many other views.

During the 2016 Presidential election, D-Tech helped distort the context in which we view candidates. The Trump campaign hired Cambridge Analytica to apply psychometric micro-targeting. Psychometrics attempts to learn and leverage aspects of personality such as psychological neuroses, sexual orientation, level of anxiety, and level of happiness. Through psychometric targeting, a campaign focuses on the combination of issues and tone to which a particular voter will most likely respond positively (Benet, n.d.).

The Cambridge Analytica company used Facebook information from 50 million Americans to develop its profiles, through which it "psychometrically micro-targeted" advertisements (Braga 2018). An anxious voter, for example, might have been provided the calming message, "Serving our families." A confident voter, on the other hand, might have been given a violent call to action around that voter's topic of interest, "Fight the immigration war!" (Benet, n.d.).

For individual voters targeted through D-Tech in this way, the 2016 Presidential election was not a choice between relatively robust views of the candidates. Rather, through psychometric micro-targeting, D-Tech created tiny, personalized windows through which voters viewed each candidate. Voter perspectives were based on partial views of the candidates.

There are caveats. Politics has been about customizing messages for years. We will never know how impactful Cambridge Analytica's micro-targeting was: to have an impact on voters, that targeted development of a warped context had to successfully recognize voters' personalities—and how those personalities would respond to a particular perspective. Such targeting is tricky (Resnick 2018). Moreover, D-Tech also helps provide holistic candidate comparisons for anyone who makes the effort to find them. Such robust comparisons can provide effective context for a prospective voter.

Nonetheless, political operatives provide most of the information on candidates that we receive. Democracy can only reflect voters' views if voters are provided a reasonably complete and shared perspective of the candidates. And, through D-Tech's micro-targeting we are eliminating that shared context. D-Tech enhances political operatives' ability to customize and twist our perspective on politics to an unprecedented level. The message each voter hears is becoming increasingly targeted to the message that voter wants to hear. We no longer receive a relatively complete perspective on an entire candidate's platform and personae. In short, we are increasingly losing the context with which we can effectively interpret individual statements about particular candidates.

We conclude this section with the third challenge.

GOVERNMENT ANALYSIS OF OUR OPTIONS

Once Government has gathered information and put that information into context, Government must decide what to do. Such decision must be based on analyzing how particular actions will impact society. As we discussed in Chapter 2, analyzing the impact of any decision is highly complex today. We have vast numbers of rapidly changing options—and many decisions are interrelated to one another. Consider, for example, our public works. On the next page, Figure 4.3, are maps of our electricity and pipeline infrastructure (Meko 2016).

American infrastructure may have historically been composed of small, independent utility centers. No longer, however. Millions of previously individual pieces have been cobbled together into intertwined, multi-dimensional puzzles. Forecasting how any particular legislative or other Governmental action will impact such complexity is excruciatingly challenging.

Meanwhile, we want more—and more complex—support from our public works. Consider Fred. Fred's my car. "Hi, Fred." He wants to drive himself, but needs a more sophisticated public works infrastructure to do so. One view of what Fred might need in the city of Atlanta is:

- 50,000 environmental sensors
- 20,000 pedestrian and mobility sensors
- 10,000 cameras

These requirements are in addition to the existing 960 traffic lights and more than 50,000 streetlights. All of these public works are needed to "up-convert [Atlanta's] city grid to so-called 'smart street' status (Padgett 2017)." Of course, this is our view today of what Fred will need. Technology is changing fast. Tomorrow, Fred's needs could appear materially different. What happens if the lighting is over-upgraded, but fewer sensors are installed? I don't know—but I do know those decisions are likely interrelated.

U.S. Electric Infrastructure

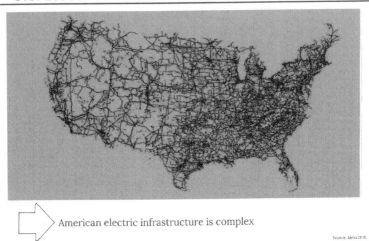

American electric infrastructure is complex

Source: Melo 2016.

U.S. Pipeline Infrastructure

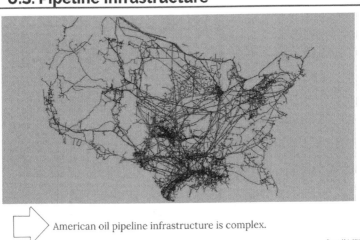

American oil pipeline infrastructure is complex.

Source: Melo 2016.

Figures 4.3: American public works
infrastructure is spectacularly complex.

Our public works needs are ever changing, ever increasing, and ever more complex. Such complexity stresses Government's ability to fulfill its role. As D-Tech increases what we can do and want to do, our society becomes more challenging to govern.

In short, through D-Tech we have created a challenging environment for Government. Our Government must now gather global, complex information to regulate our nation; it must speak many languages. It must place that information into appropriate context, despite increasing challenges obtaining un-manipulated clarity about that context. And, it must analyze the potential impact of decisions in our spectacularly complex, interconnected world.

HOW HAS GOVERNMENT ADAPTED TO OUR D-TECH ENVIRONMENT?

Poorly. We gave Richard a stash of martinis and then asked him to run our country. What did we expect?

We have many dedicated public servants. Yet, our Government is virtually devoid of leadership. As Olivia Solon (2017) describes:

> *In the 20th Century, politics was a battleground between grand visions about the future of humankind. The visions were grounded in the Industrial Revolution and the big question was what to do with new technologies like electricity, trains, and radio... Today, nobody in politics has any kind of vision; technology is moving too fast, and the political system is unable to make sense of it.*

Government has crippled its leaders. It is virtually impossible for these leaders to develop insight about our world and how it works. They are limited in their ability to take detailed actions that would make any coherent vision real. People are in charge of our Government, but few are providing true leadership. And without true leadership, we no longer respect those in charge. Often, we despise them.

Our Government leaders lost their insight about how the world works some time around 1979. Throughout the 1970s, Government exerted increasingly valiant efforts to maintain its understanding of the world. To maintain their understanding, Congress almost tripled its policy support staff from 1,669 in 1970 to 4,377 in 1979. It also increased the number of Congressional committee meetings to discuss and debate policy: during that period, these meetings grew from 8,500 in 1970 to more than 12,000 in 1979 (Drutman and Teles 2015).

Such an approach could not continue for long into the D-Tech era, however. D-Tech continued to make society more complex and continued to accelerate change. At the 1970s growth rate, by 2020 we'd need a massive stadium to hold the 200,000 Congressional staffers and their almost 50,000 Congressional meetings.

In 1979, the U.S. Government essentially gave up trying to understand our world. It cut the number of staff positions, just as the number of different specialty professions was beginning to explode across the rest of the U.S. Since then, there has been a steady decline in the U.S. Government's "capacity for generating and acquiring information...Today, the [Government agencies that] provide nonpartisan policy and program analysis to lawmakers employ 20 percent fewer staffers than they did in 1979. . . With few exceptions, our policymaking institutions are losing the ability to think for themselves (Drutman and Teles 2015)."

Figure 4.4: With today's D-Tech complexity, we would need a stadium full of staffers for Government employees to fully inform politicians. As the world has become more complex, maintaining such a staff has become impossible.

Special interests were happy to replace neutral analysts. In the place of such neutral analysis, our leaders now receive fragmented, special-interest-infused perspectives:

> *The number of organizations with Washington representation more than doubled between 1981 and 2006, from 6,681 to 13,776... The number of Washington-based think tanks more than tripled... from 100 to 306... and lobbying expenditures [grew] even more, from an estimated $200 million in 1983 to $3.24 billion in 2013 (Drutman and Teles 2015).*

These people now providing Government with perspective are responding to both the gap in Government perspective and to increasing business consolidation. As we will discuss in the next chapter, more and more business is being conducted by organizations large enough to have the money to influence our Government.

Moreover, as we discussed in the introduction, these powerful organizations *must* be present in Government. With many of our rising number of regulations broadly and imprecisely worded, the average professional *accidentally* commits three felonies a day. Any attempts to deprive someone who hires you of "your honest services" (whatever that means) can be prosecuted as a felony. Any actions that might support charges of "breach of peace," "false statements," or "obstructing the mails," for example, could be similarly prosecuted (Silverglate 2011, 35–52). Dr. Peter Gleason, a psychiatrist who specialized in sleep disorders prescribed medication for fellow physicians seemingly allowed by law—but was bizarrely prosecuted for conspiracy. Even a high profile conviction of Arthur Anderson, LLC was based on doing something many of us do every day: following the normal document retention and destruction policy before receiving a subpoena (Silverglate 2011, 35–52).

In each area of society, the relevant broad and imprecisely worded regulation differs. Yet, they're there—almost everywhere. In such an environment, large organizations must interact with the Government to minimize the risk of capricious prosecution, lawsuits, or other actions.

The people representing such special interest have little desire to help our leaders develop inspiring visions for the future based on solid, neutral analysis. These special interests want policies that help *them*.[6] Meanwhile, our leaders' ability to actually deliver any envisioned change is also neutered. Executing on any coherent vision would require passing multiple related laws. However, increasing political polarization is making the passage of any one law, much less a raft of related policies, increasingly challenging. There are multiple reasons for this polarization, but central is the role of information bubbles in which we are each cocooned. As we discussed in the last chapter, D-Tech delivers news consistent with the beliefs of the party with which we affiliate, thus pulling us apart from those of the other party. Democrats will increasingly work only with Democrats, and Republicans with Republicans (Andris et al. 2015). For example, if there was no polarization, any policy could achieve the 50 votes necessary for a Senate majority from either party. It could attract any of the 100 votes that are cast. Today, proposed policies are realistically seeking votes from one party only; it must attract virtually every single one of the roughly 50 Congressional votes the proposing party has. That's challenging to achieve.

D-Tech has also, indirectly, led to challenges controlling the execution of any legislation that does pass. By the 1970s and 1980s, explaining complex tradeoffs to voters was extremely challenging. Legislators needed laws that would allow them to take credit for lofty achievements while blaming bureaucrats for the unpopular regulations that delivered those achievements.[7]

Meanwhile, more decisions needed coordination across states. Legislators responded with national legislation that set lofty goals. The downside of such legislation, however, was that responsibility was passed from elected Government representatives to unelected bureaucrats.

The Clean Air Act of 1970, for example, requires the EPA (Environmental Protection Agency) to maintain clean air standards "based on the latest science." "Clean Air" is a popular, lofty goal, for which legislators are happy to take credit. Achieving this goal requires flexibility to change regulation with the "latest science,"[8] and needs cross-state cooperation. The Clean Air Act promotes both this flexibility and coordination.[9]

Meanwhile, legislators are insulated from the cost of achieving these goals because bureaucrats, not legislators, design the goal-achieving regulations. When those regulations such as shutting down coal-powered power plants are unpopular, legislators simply blame the bureaucrats (Schoenbrod 2017). In the face of the complexity of today's world, this legislative design sounds wonderful. "The system [has become] so politically profitable that politicians from both parties [have shown] practically limitless enthusiasm for giving citizens rights to protection (Schoenbrod 2017)."

Unfortunately, such legislation comes with at least one downside: execution is difficult for today's elected Government representatives to control.[10] Legally, bureaucrats are *required* to achieve the lofty goals. These bureaucrats *must*, for example, implement any clean air regulation that—viewed independently of anything else—has benefits greater than its costs.

However, what happens when such regulations have indirect impacts? What happens if multiple regulations are inconsistent with each other? What if satisfying one regulation makes another regulation more challenging with which to comply? What if a better structured set of clean air regulations would both clean air better *and* reduce compliance costs?

Nothing happens. (At least, nothing straightforward happens.) The lofty goals written into the 1970s and 1980s regulations say nothing about indirect impacts. They say nothing about considering multiple regulations at once. As a result, our masses of individual regulations end up looking like our electricity infrastructure: a confusing labyrinth of interconnected spaghetti. Complying with such a labyrinth of regulations is exceedingly expensive—and it protects us less than it should.

Foresight is challenging. It may have been impossible to foresee in the 1970s the challenges created. However, revising legislation is also challenging. Legislators could likely write lofty goals more appropriate to today. The 2017 Federal Automated Vehicle Policy, for example, outlines a creative new approach to regulation design (Department of Transportation 2018). In that 2017 policy, regulators set goals and allow Businesses to define how they will achieve them.[11] How will a self-driving golf cart ensure it's safe? Most likely differently than a self-driving forklift will.

In principle, legislators could replace old legislation with new legislation updated to target more refined goals. However, passing any new legislation is challenging.[12] And, such refinements today cannot make the changes we will only realize we need in another 30 years.

Our D-Tech Culture has also had an impact. It has lowered our standards for what we will accept. Or, perhaps we've just accepted reality. Either way, we now expect less from our Government leaders. Fewer and fewer people expect Government officials to act ethically. Perhaps we understand the challenges providing inspiring leadership today. Perhaps we are sickened by reports of unethical, or potentially unethical activities. Whatever the cause, decreasing numbers of citizens view our judges, members of Congress, and governors as having high ethical standards (Brenan 2017).[13] While they do not have to be, such low expectations have a history of being self-fulfilling.

In short, our Government has responded to the challenges of D-Tech poorly. It has compounded the problems. D-Tech has made our environment more challenging to understand. Government has given up trying to maintain a strong cadre of neutral Government expertise. It has allowed special interests to fill the knowledge void.

D-Tech has increased society's polarization. D-Tech has made it more challenging to explain difficult tradeoffs to voters. Government has responded by accepting new limitations on what it can do, and has passed legislation that must be implemented by bureaucrats, rather than by the legislators themselves. We no longer expect leadership, at least ethical leadership, from those in our Government.

Digital Technology is not directly at fault: there's no computer villain who secretly threatens every American president. Nonetheless, D-Tech is indirectly to blame. D-Tech changed the environment in which we live and work, and these changes led to the current state of our Government. In many ways, D-Tech, not our leaders, controls Government.

WHAT IS THE IMPACT OF GOVERNMENT ADAPTATION?

Our Government isn't setting the stage we need.

Government plays important roles in society: it prepares the venue in which everyone else in society operates. Government Regulations determine the edge of what the rest of society can do. For example, Government rules support contracts and property rights while constraining our most destructive impulses such as fraud and murder. Government's public works (e.g., electric supply and roads) and shared services (e.g., military) are the props that determine what is easy for the rest of society to do. Our safety and access to stable electricity and internet, for example, make it easy for us to open an internet-driven, website design business.

In addition, Government's taxation and subsidization pays for Government's activities. Taxes also transfer money to targeted groups and individuals. Huge numbers of our elderly, for example, now rely on payments from Government's social security program.

How is our Government stagehand setting the venue for today's stage?

The stagehand was Richard. He puked on stage.

Government expends vast resources: In 2011, American Government spent an amount equivalent to roughly 35–41% of the value of everything produced in the U.S. (Ruffing 2011). Government employs 2.8 million federal employees ("United States Federal Civil Service" 2018).

But what does their work look like? Put 2.8 million people together in a massive concert hall, and tell them to work diligently to prepare the venue for the evening's drama. Then, increase the challenge further by giving them ineffective leadership.

Such a mass of humanity fails to prepare an orchestra pit for inspiring music: they place unaligned scores in front of the percussion and woodwinds. They crack the clarinet's mouthpiece so it squeaks. They tear the drums' wraps, and they squirt water in the tuba.

Without leadership, Government obstructs rather than facilitates. Today, we face angry seas of expensive regulation. These regulations are excessive and too frequently special-interest driven, yet also incomplete. Our public works are unstable and unsatisfying. And, we are consistently expanding our public debt.

Figure 4.5: The U.S. Federal Government Band
is unaligned and their instruments need tuning,
yet they are still playing their hearts out.

REGULATIONS

How did we end up with regulations that are excessive and too frequently special-interest driven, yet also incomplete? In September 2010, then Minority Leader John Boehner bemoaned the state of policymaking: "The institution does not function, does not deliberate, and seems incapable of acting on the will of the people. From the floor to the committee level, the integrity of the House has been compromised. The battle of ideas—the very lifeblood of the House—is virtually nonexistent (Drutman and Teles 2015)." While John Boehner was specifically speaking about the House of Representatives, the same could be said about much of Government.

Perhaps Government is simply too busy creating legislation and regulations to deliberate and think. As shown in Figure 4.6 [14] in 1970, we could print every federal regulation on 55,000 pieces of paper. By 2013, it took 170,000 pages: regulation more than tripled in 43 years. With D-Tech, it's easier than ever to research and write new legislation—and this simplicity is illustrated in the volume of current legislation.

With D-Tech, it's also straightforward to create new Government agencies—and the number of regulatory agencies providing oversight exploded with the numbers of regulations: The Bureau of Ocean Energy Management, Regulation, and Enforcement, The Fiscal Service, Federal Financing Bank, Office of Financial Research, and the Office of Foreign Asset Control. In 1970, roughly 100 federal Government Agencies existed. That is only 40% of today's 250.

The Government doesn't oversee 250 distinct areas, however. At last count, 16 different agencies oversee intelligence ("United States Intelligence Community" 2018). Ten different offices run AIDS programs for minority communities, and autism research is spread over 11 agencies (Korte 2014). Such overlap is the norm, not the exception.

U.S. Annual Code of Federal Regulation Pages

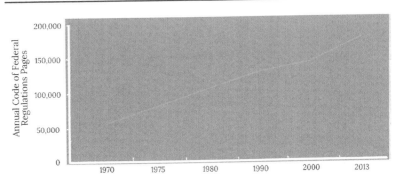

U.S. Federal regulations have expanded rapidly throughout the Digital Technology Revolution.

Source: "Code of Regulation Pages" (n.d.) and Crews (2014)

Figure 4.6: The number of pages of federal regulations has more than tripled since 1970–increasing by tens of thousands of pages under every President.

Are you concerned some of these agency employees are bored? No need. Each agency that existed in 1970 only oversaw 0.4 million restrictions (see note 1). Today, as illustrated in Figure 4.7, each agency oversees more than 1 million restrictions. Every agency is hoping that its particular instrument will be heard, independent of whether the result is a beautiful symphony—or a cacophony of discord.

The result of such confusion is ugly—and expensive. Regulations enacted since 1980 cost every man, woman, and child in the United States an estimated $13,000 to $30,000 each year (between 25% and 50% of the U.S. economy).[15]

U.S. Federal Agencies and Restrictions per Agency

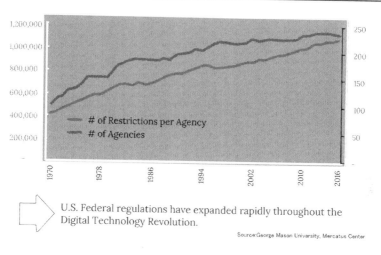

U.S. Federal regulations have expanded rapidly throughout the Digital Technology Revolution.

Source: George Mason University, Mercatus Center

Figure 4.7: The complexity of the American federal government has exploded with D-Tech. Both the number of federal agencies and the number of rules each agency enforces have expanded (Coffey, McLaughlin, and Peretto 2016).

Some of the cost is the result of specific laws and regulations. In the face of D-Tech complexity, even straightforward destructive impacts can "hide in plain sight." One law blocks the U.S. Government from negotiating the price of drugs for the elderly who participate in the Government-operated Medicare system ("Subpart 2—Prescription Drug Plans," n.d.). The impact of such regulation on what used to be "negotiation" is clear: the pharmaceutical salesperson tells the Government buyer, "Next year, we'd like to charge you 28% more."

Most Governments outside the U.S. negotiate pharmaceutical prices. The U.S. Government buyer responds, "Yes, please increase price. Are you sure you don't want more?" The pharma sales reps certainly owe their lobbyist for that.

In the face of D-Tech-driven complexity, we find that other regulations are contradictory. We spend $322 billion treating diabetes, much of that coming from the Government (American Diabetes Association, n.d.). However, we take few steps to limit the sugar consumption that helps cause the disease, for example (Bunim 2013).

At least as troubling as individual regulations, however, is the combined mass of regulations. "Each agency operates as if it is the only one issuing regulations. . . (Dunkelberg 2016)," solving its own targeted problems. As a result, many regulations are "duplicative and conflicting (Coffey, McLaughlin, and Peretto 2016)." The Holland Tunnel in New York, for example, needs a facelift. To date, according to the Wall Street Journal, this retrofit has received 19 permits from 29 participating government agencies (Editorial Board 2017). The retrofit is currently expected to cost $13 billion over 8 years. That's more than 20 times what the tunnel cost to build originally (over 7 years) even after adjusting for the rising cost of material and labor.[16]

It's more than just the Holland Tunnel. Even a traditional highway permitting process is challenging. Such a project requires 16 different approvals involving 10 different federal agencies being governed by 26 different statutes (Trump 2017).

These are the angry seas of regulation. This regulation actively obliterates constructive interaction in wide swaths of society.

In a different environment, some of our crazy regulations could make sense.[17] And, yes, along with the overwhelming cost of such regulations also come benefits. There are fewer toxic chemicals in infant formula impacting the brain function of our infants. Our cars are likely safer, and they emit less pollution. We might even be better off with all of the regulations that exist today than if we eliminated them all—despite the suffocating cost.

Nonetheless, such benefits cannot excuse the angry menace that regulations pose today. Wading through such masses of rules sucks up resources while spitting on creativity. And, we become frustrated and confused. We're unable to even think about what to do next.

Just as vexing as the other regulatory problems are the gaps in regulation. We have important areas within which the limits of acceptable behavior are undefined. These regulatory gaps are particularly common (and problematic) in virtual areas we are newly cultivating. Some of these gaps arise from the borderless nature of our virtual world.[18] Others arise from the unique nature of interactions in the digital world.

Governments may be constrained by physical borders, but their responsibilities now frequently cross borders (Bratton 2016). Without clear regulation, the Government is frequently confused. "Say you sign up for a free email account. You live in country A, the tech company providing the email service is headquartered in country B, and the data center where the emails are stored is in country C. If the police in one country want your emails, where do they have to go, and whom do they have to ask (Jeong 2018)?".

Business is similarly confused by the regulatory requirements for handling digital information: "What should companies do if the privacy standards in the country where they are holding the data are far more stringent than those of the country asking for the data? Where there's a conflict of laws, must and should companies give governments information for investigations of things that are legal in the U.S. like homosexuality (illegal in Bangladesh), talking trash about the king (illegal in Thailand), and holocaust denial (illegal in France) (Granick 2016)?"

The European Union Government has clarified its perspective: "the European Union's General Data Protection Regulation

(GDPR) explicitly seeks to apply EU data privacy principles to data controllers processing the personal data of Europeans regardless of whether the processing itself takes place within the Union or not (Harkness and Jaffe 2016)." What do such regulations imply for every American and every American business operating in the United States? We're still unsure.

We equally have material concerns when there is a combination of physical and digital interaction. Consider, for example, the practice of brushing (Shepard 2017) that crosses American and Chinese borders. "A Chinese company sends one of its inexpensive products, such as a plastic hairband, to an American consumer. Using the consumer's name, the company then posts a fraudulent review online—for some expensive product. The worthless hair band shipment is used as proof that the consumer is a 'confirmed' customer."

If all of the activities occurred in either China or the United States, brushing would qualify as fraud. However, two countries are involved. The connection is virtual. As a result, Chinese companies profitably "brush" with little fear of punishment.

For many years, technology has shortened what historically has been long distances. As transportation improved throughout the era of Industrial Technology, people could travel first by clipper ships and trains, and later by cars and planes. Governments have increasingly had to compete with each other on business tax rates, regulations, and others, as a result.

With virtual technology, this elimination of distance becomes complete: in our virtual world, location is meaningless. How we respond to this lack of meaningful distance in our virtual world, which conflicts with the traditional significance of physical distance, will continue ... until we clarify such distinctions.

Even within the borders of one country, our virtual world creates a host of new challenges for which effective Regulation is thin or absent. Consider, for example, the largely virtual companies, Google and Facebook. What is your deepest, darkest fear related to those companies? Perhaps it is one of these possibilities:

- **Culture Manipulation:** These companies have powerful influence on society. What happens as the number of positive and negative words is adjusted in Facebook's news feed? User mood seems to change—if only (thus far) by a modest amount (Meyer 2014). Meanwhile, Google's search algorithms have identified pictures of two black people as "gorillas." Perhaps this result is best summarized by the message the people identified as gorillas sent back to Google: "Google Photos, y'all f**ked up" (Grush 2015). Whether they f**ked up or not, effectively calling black people "gorillas" has a strong influence on our culture. It reinforces and redefines our stereotypes.[19] What *must* these companies do to respect our culture?

- **Out of Control AI:** Artificial Intelligence (AI) is incredibly powerful technology, and Google and Facebook are well known for having some of the most sophisticated AI labs. A panel of top technologists at the Asilomar Artificial Intelligence Safety Conference ("Asilomar AI Principles," n.d.) concluded that it is likely that someone will create a self-developing artificial intelligence. In other words, it's likely someone, at some point in time, will create software able to get smarter and smarter and smarter on its own. And, if anyone does create such an AI, Facebook and Google are reasonable bets. Are there any legal limits to what these companies can do?[20]

- **Loss of Privacy:** Information is regularly stolen from the virtual world. Recently, for example, the company Ashley Madison (unassociated with Facebook and Google) had information on thousands of users stolen. Ashley Madison "enables extramarital affairs" (Chappell 2015). Yes, people were embarrassed (and worse) when they were exposed ("Ashley Madison Data Breach" 2017). Less sexy, but equally important, is the status of our personal medical information. D-Tech creates value by using information at multiple locations—but every location that has access to our information is also a security concern. What exactly are the expectations for privacy that consumers can expect? And, what rights do consumers have over their own information? What is the equivalent of property rights in the virtual world, and how important are they?[21]

Each of these areas is a regulatory desert. There are promises to develop recommendations on how to handle such questions.[22] There is increasing legislation and judicial precedent in some of these areas. However, none are deeply developed (Microsoft 2018). And, clarity about the "rules of the game" in each is critical for developing our virtual world (Delaney 2018).

It could be tempting to consider the benefits of a virtual world completely "unencumbered" by regulations. The so-called virtual currency, Bitcoin, for example, is currently free from regulation. It grew from absolutely nothing to a value of almost one-third of a trillion dollars in less than 8 years.[23]

However, "standard shenanigans" that "harm retail customers" are common with Bitcoin (Gerard 2017). These shenanigans and their results that are allegedly common in Bitcoin have names like, "flash crashes, delayed settlements, spoofing, hacks, internal theft . . . and bucket shop arrangements (Commodity Futures Trading Commission 2017)." These items harm the ability for many businesses and individuals to use Bitcoin constructively.

On the other hand, it would be easy to focus on the fear of over-regulation. As we discussed just a few pages ago, much of our physical world is over-regulated. However, fear of over-regulation does not excuse under-regulation. There's a parallel in our physical world: we do not want Government to require us to look into each other's eyes when shaking hands—that's many steps too far. However, we support laws that forbid rape, murder, and terrorist activities—with good reason. Such laws provide borders within which we know everyone *must* stay—they help curb our most animalistic tendencies. Curbing such tendencies is a powerful support for nourishing our constructive opportunities. Regulation of our virtual world must be similarly constrained.

Government moves slowly, and that is by design: better to "measure twice and cut once" as the saying goes. However, Government's current pace of change is slower than the pace of societal change. We will continue to find ever more areas of new endeavor—areas within which destructive tendencies need to be constrained. Government lagging far behind, covering areas of regulatory interest with thick swathes of regulation—well after the fact—cannot be acceptable.

PUBLIC WORKS[24]

The water disaster in Flint, Michigan puts a human face on public works challenges in the United States:

"Seemingly out of nowhere, 9-year-old Nicholas Carr will become violently ill. His stomach will hurt and then he will vomit. This never happened before lead leached into the [Flint, Michigan] water supply (Adams, n.d.)." In the most affected Flint neighborhoods, 10% of the children have elevated lead levels in their bloodstream (Torrice 2016). Across Flint, birth rates during the period of contamination were 12% lower, and fetal death rates higher. In the future, Flint will likely face poor academic achievement and increased crime as kids Nicholas' age grow up (Economist 2017).

The problem with our public works goes well beyond Flint's water, however. The American Society of Civil Engineers gives American infrastructure a grade of D+ for 2017 (American Society of Civil Engineers, 2017a). "Nearly half of all vessels [using our inland waterways] experience delays (American Society of Civil Engineers, 2017c);" "24 of the top 30 major US airports may soon experience 'Thanksgiving peak traffic volume' at least one day every week (American Society of Civil Engineers, 2017b);" and "More than 2 out of every 5 miles of America's urban interstates are congested and traffic delays cost the country $160 billion in wasted time and fuel in 2014 (American Society of Civil Engineers, 2017d)."

American public works are insufficiently targeted, unstable, and unaligned with current needs. Joel Moser (2017) illustrates the importance of leadership vision for public works through a transportation-focused discussion. A true leader's vision must guide us as we answer question such as:

> Will we build roads or rail? Or more specifically, will we enable more cars or encourage mass transit? In rail, are we building more commuter trains or inner-city systems? Basically, are we facilitating more urban lifestyles or suburban communities? Do we care about tailpipe emissions? Which begs the question of whether we accept the science that humans are a cause of climate change. Or do we believe that we are fast approaching an electric-car future charged by renewable power?
>
> In longer haul, domestic transport, are we building high-speed rails or more runways for regional flights? Sorry we're back to climate change, but are we confident about jet biofuel and solar-powered aircraft? What about hyper-loops instead?

With D-Tech obscuring Government's ability to develop a clear vision, the current mediocrity of our public works can be expected. The complexity accompanying our use of D-Tech is blinding Government's ability to clearly see any details. The resulting lack of vision is becoming all too clear in the complex puzzle of interconnected, puzzle-like elements in our public works. Our investment in infrastructure is being stretched too thinly. We can certainly spend more and perhaps we'd even obtain some benefit from that investment. Yet, the question is the focus. Is our priority a wonderful urban setting? Or, perhaps it's facilitating wonderful suburban homes with white picket fences. Instead of a handful of world-class, integrated public works systems—we have a mediocre version of every public works class that exists.

Such increasingly complex public works systems are also unstable. Flint water works was mismanaged. The impact of that mismanagement was magnified by the instability in Flint's legacy water system, however. The devastating flow of poison coming from Flint's water taps was the result of both. As Flint's history unfolded, different raw materials were used for pipes, including lead, copper, and steel. Each of these materials reacted differently with river water. The result was a complex, difficult-to-predict system.

In such a complex system, almost any change could cause a cascade of consequences, some of which threatened to be disastrous. Figure 4.8 illustrates the cascade that may have led to Flint's poisoning: at least *some* part of Flint's water system reacted differently than anticipated to *every* decision.

Most American water systems are similarly unstable. Most water systems have complex legacies. There are millions of lead pipes in water works across the United States (Torrice 2016). There are millions of steel pipes. We install D-Tech systems to help us manage those systems, and those systems do help.

However, those systems do not help us wrap our human minds around the complex interplay of our public works systems. We simply expect, and hope, that most water managers are more aware of how their systems are evolving than were Flint's.

The instability of American Public Works goes far beyond just water systems, however. As we apply Digital Technology, we are increasing the complexity of all public works. Most of American public works are the results of complex legacies. Different investment decisions are made at different times, and the results of those decisions are cobbled together with the help of D-Tech. In the early 20th Century, electricity was supplied by small, local facilities. As technology improved and we moved into the Digital Technology era, those local supply systems were linked into increasingly large ones. As a result, "In the 1950s through 1980s, significant power outages averaged fewer than 5 per year . . . In 2007, there were 76, in 2011 more than 300 (Davies 2016)." In 2015, the U.S. suffered 941 outages caused by faulty equipment and human error (Eaton, n.d.).[25] Increasingly, cascades of unintended consequences are the likely result of any change to our public works, and change is accelerating. Similar outcomes can be described for our internet, development of traffic jams, and more.

Even complex, unstable public works can generally be managed effectively and safely. Thankfully, the lead poisoning in Flint Michigan is the exception, not the rule. However, this legacy-induced instability certainly decreases the quality of our public works. If we could simply start over, the decisions made would be different and the resulting systems far simpler and more robust. We can't. As we apply D-Tech to create an increasingly complex world, our public works are likely to become increasingly unstable.

Flint River Disaster Scenario*

The Flint water disaster resulted from a cascade of management decisions and consequences.

* Note: It is uncertain whether this specific cascade occurred. It is one of several possible.
Source: https://com.ucs.org/articles/94/G/Lead-Ended-Flints-Tap-Water.html

Figure 4.8: A complex series of events preceded the lead poisoning of thousands of Flint, Michigan citizens. Preventing that disaster would have required managers understanding how the complex system was evolving (Torrice 2016).

We are still able to use our public works. However, as change across society accelerates, new technology demands more and more from our public works. This demand is increasingly unaligned with our public works' ability to satisfy that hunger.

In short, our public works are critical to our collaborating effectively. Yet, they are also insufficiently targeted, unstable, and unaligned with current needs. We may feel comfortable in our homes and offices on a daily basis. However, when we look even slightly below the surface, we find problems. Our Public Works earn that grade of "D+."

TAXATION AND SPENDING

The current Government deficit is unique to American history: it is the first, sustained, spending deficit *not* associated with a major war.

Look at the history of American Government spending in Figure 4.9. You can easily pick out the past major wars. To pay for each major war, Government debt spikes. That debt is then paid off during the 30–40 years after the war ends. Such debt peaks existed for the War for Independence, the Civil War, World War I, and World War II (Agresti 2018). We stopped paying off the World War II debt spike in the 1970s.

National Debt as a Portion of the U.S. Economy, 1790-2016

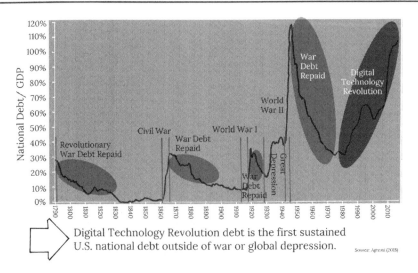

Digital Technology Revolution debt is the first sustained U.S. national debt outside of war or global depression.

Source: Agresti (2018)

Figure 4.9: The D-Tech era is the first extended period of large, expanding federal debt outside of a major war or massive economic depression.

For the 35+ years since the early 1980s, we've created yet another Government debt spike. Current debt is close to 110% of total economic production (GDP). This spike is almost the same size as the World War II debt spike (which peaked at less than 120% of GDP.) In other words, when viewed from the perspective of Government spending—the Digital Technology era is equivalent to World War II.[26] Eliminating the current deficit is not a trivial adjustment exercise. In 2016, the Government spent 21% more than it brought in as taxes and other receipts.[27] For an individual earning $100,000 per year, that's equivalent to spending $121,000. Government receipts didn't even fully pay for the top four spending categories: healthcare, pensions (including social security), interest on our debt, and military spending. Each of these four categories of spending is likely to grow, not shrink, if current policy continues. In other words, even in the unlikely event *everything* else Government pays for is eliminated, Government debt would continue to increase. Of course, this everything else includes agricultural subsidies, education, international affairs, scientific research, and more.

Because the U.S. economy is so large, we can handle the amount of debt we currently have.[28] Relative to the size of the economy at the time, we managed moderately more debt after World War II, after all. Our total current debt is similar in magnitude to someone with an $80,000 a year income having $88,000 in credit card debt. Such debt can be paid off, but there are sacrifices associated with doing so.[29]

There are attractive futures in which U.S. Government debt is paid, even with only moderate sacrifice. Over the course of this book, for example, we outline a number of societal challenges. Each of these challenges is massively destructive to our economy. Increase the size of the economy, and the debt becomes easier to pay.

Nonetheless, the uniqueness of our situation is disturbing. With today's D-Tech-supported complexity, even clearly explaining challenging tradeoffs to voters is difficult to do. It appears our Government representatives have stopped trying to explain the tradeoffs needed to balance the Government budget. We have the first massive Government debt not associated with a major war. During the Great Depression, the U.S. Government did borrow about two-thirds of the amount used during the major wars. However, we're not in a depression—but we *are* in a 35+ year spending spree.

It's possible that our current debt is not caused by the D-Tech-driven change in our environment. Perhaps our expanding debt is due to a change in our culture, rather than due to increasing complexity. We may now want to consume everything today, our children be damned. That would be sad.

We can find solace in not having the loss of life that World War II caused. However, we face equivalent debt.

SUMMARY: IMPACT OF DIGITAL TECHNOLOGY ON OUR GOVERNMENT

Digital Technology has transformed our Government. Government has clearly benefited from applying Digital Technology. Every law is recorded and accessible digitally. Every Government representative can now communicate via email.

However, D-Tech has also *indirectly* impacted our Government, as we face a litany of emergent impacts. These indirect impacts of D-Tech have been confusing and destructive. To manage the complexity created by D-Tech, we've created a whopping 250 federal Government agencies. Many of these agencies—individually—employ more people, and conduct more varied activities than do many of our largest companies.

Meanwhile, it is increasingly challenging for Government to think clearly. Government must now gather global information to regulate our nation. It must now understand the specialty languages of many industries, cultures, and technologies. Government must analyze complex, interconnected systems that look like nothing more than jumbles of spaghetti to the uninitiated. In fact, many of these systems even look like jumbles of spaghetti to those who are experts. And, Government must operate as a democracy despite increasing potential for voters to be manipulated. In other words, Government must operate in a maelstrom of noise and confusion.

In many ways, D-Tech's maelstrom controls our Government, and our human leaders do not. No longer can our human Government leaders expect to consistently develop deep insights about our world based on neutral expertise; they are dependent on special interests. No longer are these leaders fully capable of developing and implementing a clear vision for our future and controlling that vision as it unfolds.

As a result, our regulations are too frequently special-interest driven and excessive, yet also incomplete. Our public works are unstable and unsatisfying. For the first time in our history, we are spending as if we were fighting a world war—but the only major war is with D-Tech. We are constantly making improvements and reacting to the flaws in our system as it is. And we are always two steps behind.

We are unlike the successful Agricultural Revolution societies. We are not progressively evolving our societal institutions in the face of increasing challenges. We are following the path of the failed Agricultural Revolution societies. Our Government is drowning in potential.

In Chapter 5, we will complete our exploration of the impact of D-Tech on our society. Are the benefits we obtain from

Business and our Economy progressing, as that sub-system helps us leverage D-Tech? Or, is this system becoming overwhelmed by D-Tech, like our Culture and Government? Let's find out.

CHAPTER 5

THE IMPACT ON AMERICAN BUSINESS:

Economic Perversion and Our
Critical Business-Skill Failures

In this chapter, we look at the impact of Digital Technology (D-Tech) on our economy. This perspective builds on my Ph.D. in economics from the University of Michigan, as well as my experience as a business leader and management consultant with startups through Fortune 500 companies.

D-Tech has allowed us to create amazing new surgical tools, teaching aids, and efficient factories. Such creation is wonderful.

Yet, our businesses and economy are also failing to serve society along too many critical dimensions. In this chapter, we investigate the relationship between D-Tech-driven complexity and these failures. We regularly introduce amazing new products, and profits have recently been at record levels. At the same time, we face *unaddressed needs* as a society: more than 60% of our economy is dedicated to industries failing to help us obtain our most fundamental, basic needs. For example:

- We see a constant inflow of amazing, new healthcare products. Yet, we are not efficiently maintaining good health.

The per-person cost of U.S. healthcare was 68 times larger in 2015 ($9,990 per person per year) than it was in 1960 ($146) (Amadeo 2017).[1] We spend more on healthcare than do the people in any other developed country—yet our lifespans are the shortest.

- Our financial markets provide us spectacular financial instruments. Yet, the prices of those financial instruments are too frequently associated with terms like "bubbles," "cascades," and "instability." People outside the financial markets make decisions based on such untrustworthy prices—which may not be in society's best interest, as a result.

- The average supermarket today offers more than 47,000 products (Consumer Reports 2014). Yet, those offerings are not nourishing us. Even our organic foods of today are less nutritious than their traditionally farmed counterparts from the 1950s and 60s (Plumer 2015).

What caused such failures? As we see in this chapter, we changed the core economic problem in three ways. First, deciding what to produce used to be the most challenging problem. Do I grow wheat or sorghum, and how much. Price helped us coordinate: the higher the price, the more I should produce.

Today, we can modify and track the world around us in many powerful ways. Our greatest challenge is coordinating *how* we produce. If I build a Honda car body and you build a Ferrari engine, for example, numbers of engines and bodies align, yet the outcome remains poor. Questions about *coordination of how we use resources* have surpassed challenges of how we *allocate resources*.

Second, we also face more questions about how to describe to consumers what we produce. Wheat is no longer "wheat." I can now adjust how rapidly your stomach digests wheat, the amount of vitamin B9, and potentially the molecular structure of that B9. Clearly explaining to non-specialists what is different about such wheat is challenging.

Third, our digital machines, which have intelligence in only very narrow areas, can now impact how our relatively rich human minds think. How do we control such impact?

As we will see, we are responding to such challenges poorly.

In less than 50 years, the impact of D-Tech on Business has grown from a small eddy helping Business progress to a powerful undercurrent. D-Tech drives Business. Yet, in too many cases, D-Tech's *side effects* also confuse us, sicken us, and squander our resources.

Our experience with D-Tech contrasts with the experience great societies of the past have had with their technology. Thousands of years ago, for example, great leaders helped reshape Business to respond to Agricultural Technology. As Agricultural Technology was increasingly applied, ex-nomadic societies became settlements. As Agricultural Revolution settlements grew, settlements' needs became more diverse and the knowledge of its citizens became more specialized. Markets in which Business could thrive had to be developed. In successful societies, leaders reinforced the importance of currency and market prices while supporting legally binding contracts.

Today, in contrast, we are drowning in potential. We are not constructing our society to match our current challenges. Business may be delivering record profits and offering shiny, new products. However, D-Tech is perverting our Business. Every industry I analyzed, covering the majority of the U.S. economy, is failing to deliver what society needs from it. Meanwhile, much of our future is beholden to the decisions of a handful of Business leaders.

Certainly, Business benefits *directly* from D-Tech. In fact, D-Tech is critical for Business today.

Nonetheless, D-Tech's *indirect* impacts on Business are similarly powerful; we call such *indirect* or *emergent impacts* D-Tech's *side effects*. We now have powerful technology with which we can change our world. We must target that technology and coordinate the changes in our world.

I recently had a conversation with a friend of mine from grade school, Christopher. He slapped me on the back, leaving the orange dust from his processed food snacks. He went to college but never learned to interpret anything close to a complex table of nutritional information. He was recovering from a heart bypass, which he underwent for some issues related to heart pain.

> *"How's it going, Chris?"*
> *I ask. "Surf will be up tomorrow. I'm glad I still have my old surfboard. That and the Internet keep me busy."*
>
> *Chris continues, "I had to sell my parents' place right on the beach because I couldn't afford it anymore. I really thought I had something with my Bitcoin investment. I bought late, but it still more than doubled before crashing."*

As a society, we have cheap and tasty food. However, we have less and less nourishing food. We have spectacular education—yet too many children who cannot read and too many adults who cannot follow basic logic. We have glamorous financial tools—which we cannot rely on to deliver consistently appropriate prices. And we have healthcare that leaves many of us sick and others dying early. D-Tech has provided Business the tools to create spectacular, vast enterprises. Yet, it is as if the behemoth Businesses are in agony as their blood vessels work against their bones, which are working against their muscles.

Over the course of this chapter, we focus on the *second-order side effects* of D-Tech felt by Business. This discussion covers the grey-highlighted portion of Figure 5.1: as Society evolves,

Business responds to changes in the environment in which we live and work (Arrow 1). How Business evolves also depends on A) changes in the rules our Government sets for it and B) shifts in the Culture of the people participating in its markets (Arrow 2).

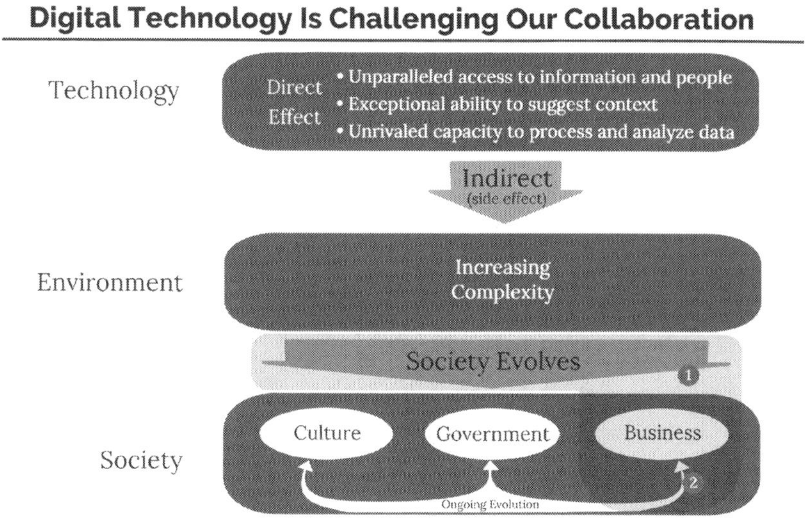

Digital Technology Is Challenging Our Collaboration

Technology — Direct Effect
• Unparalleled access to information and people
• Exceptional ability to suggest context
• Unrivaled capacity to process and analyze data

Indirect (side effect)

Environment — Increasing Complexity

Society Evolves ①

Society — Culture · Government · Business ②

Ongoing Evolution

Figure 5.1: American business and economy have evolved as a result of D-Tech induced complexity, i.e., as a result of D-Tech side effects.

So, what is the most obvious *side effect* of Digital Technology on Business? D-Tech has made it easier to provide huge quantities of a particular product or service. We can run massive factories with only a handful of workers. Once we have the software established, we can send out thousands of invoices at once. Our software platform can be accepted as the standard that everyone uses. It's moderately expensive to build one of these large factories or establish such complex software.

Yet once built, the cost of adding additional customers is tiny. You win. You get all the customers.

The "Winner-takes-all economy" (Freeland 2013) is the name of this phenomenon. If you want to participate in business, you need to either be big or get big—fast. The number of Businesses is shrinking as each Business gets bigger. The Wilshire 5000, the stock market index of the 5,000 largest publicly traded companies in the United States today, has only 3,599 firms (Wilshire 2017). There simply are no longer 5,000 sizable public firms in the United States.

Many of the remaining firms are *big*. Amazon's prime service has more than half of all American households signed up—which is more households than attend church regularly, than own a gun, or that voted for either Donald Trump or Hillary Clinton in 2016 (Stoller 2017). During its Amazon Prime Day, Amazon handles more than 400 transactions per second (Stoller 2017). Approximately 80% of seats on airplanes are sold by only 4 airlines (Creswell and Maheshwari 2017). Google and Facebook determine roughly 75% of online advertising (Ingram 2017). Four companies (ADM, Bunge, Cargill, and Dreyfus) now account for between 75% and 90% of the global grain trade (Lawrence 2011).

How many people are involved in the biggest companies? With 2.3 million employees ("Company Facts," n.d.), more people today work for Walmart than live in the city of Houston, Texas. If we put all of Walmart's employees in one place, we'd have the fourth largest city in the U.S.. Five companies employ at least a half million people—each of which hires more people than live in the 40th largest city in the U.S. (Colorado Springs, CO, with 465,000 people). In 2016, twelve companies had turnover of at least $192 billion ("Global 500," n.d.). These companies are *big*.

Not every part of the U.S. economy became big though. A growing number of U.S. employees are working in the "gig economy" (Rouse 2016) in which independent workers obtain short-term contracts for specific projects. The gig economy is currently 0.5% of the U.S. economy, but it is new and growing. And, 94% of all jobs created between 2005 and 2015 were temporary or unsteady (Katz and Krueger 2016). Some companies become dominant without being large in every way: Facebook, for example, is one of the most powerful companies on earth with only 17,048 employees (Statista 2017a). Nonetheless, *big* has increasingly become the heart of American business.

So, the largest U.S. businesses are massive. Big isn't inherently bad. What we care about are the benefits society obtains (or fails to obtain) through these massive businesses. What *side effects* are we experiencing?

We answer this question here, dividing this chapter into three sections. First, why is there a relationship between D-Tech and how well Business delivers benefits to society? Here, we expand on the description of our increasingly complex environment that we discussed in Chapter 2, focusing on those aspects relevant for Business.

Second, how has Business adapted to our complex environment? In short: destructively. Individual elements of our economy are spectacular, but they are disparate and poorly integrated. As we will discuss, Business is materially different today than it was before D-Tech.

Third, what is the result of Business adaptation? We are drowning in potential. We are getting many of the details right— but our answers remain all wrong.

I judge our economy's success on the needs its citizens are able to satisfy today, and the risks created for tomorrow. In addition to the value of what is produced in the U.S. (Gross Domestic

Product), I also evaluate the alignment between what is produced and society's needs.

Economists traditionally limit comments to the objective value of what is produced. It would be unfair to suggest the subjective "too many yellow cars are being produced" simply because I prefer blue cars. The market, as reflected in the price people pay for different products, is a much better judge of such decisions.

In today's D-Tech economy, however, misalignment between society's needs and what the economy delivers is extreme. Later in the chapter, I will illustrate that we face an expanding divergence between what we need and the financial measure of what is being delivered:

- **Education:** We have doubled spending on education per year per child with no material improvement in outcomes. Those outcomes, however, remain insufficient for consistent societal success.
- **Food:** Our food is providing decreasing level of nourishment.
- **Housing:** The gap between the housing we need in American economic hubs and the housing delivered is greater than ever before—despite more than sufficient land being available.
- **Healthcare:** Our healthcare spending has increased to current levels—twice as much per person as the average developed country—yet we are not obtaining corresponding benefits. Our life expectancy is the shortest in the developed world.
- **Finance:** The price of assets is decreasingly reflecting scarcity—causing decisions based on financial prices (e.g., price of wheat versus sorghum) to be.
- **Grey Market:** The largest share of American economic transactions ever is being made without appropriate

paperwork, eliminating the potential for government to provide its support.

- **Labor Market:** Labor force participation by prime-age males in the U.S. is the lowest ever. While participation is lower for less-educated males, labor force participation has fallen at every level of education.

These arguments cover more than 60% of the U.S. economy today—and *every sector I analyzed.*

Twisted arguments could be made that each of those outcomes is somehow efficient. Perhaps, for example, we purchase less and less healthful food because we care less and less about our own health or the health of our children. I have heard that argument.

Yet, such arguments are incredible. For example, when people receive clear information about the healthfulness of food options, they generally respond constructively. Each of the arguments made over the course of this chapter is defensible. Moreover, provided the body of evidence in this chapter, the more realistic argument is that our Business and Economy are increasingly failing to deliver what society needs and values. Our societal systems need to respond to a new set of economic problems by emphasizing collaboration tools that fit today's world.

Let's begin by discussing how Business, through our Economy, is failing to deliver what we need today.

WHY IS D-TECH STRESSING SOCIETY'S ABILITY TO BENEFIT FROM BUSINESS?

D-Tech has changed our environment. In Chapter 2, we high-lighted how D-Tech is accelerating all technological developmen. Want to design a new type of motor? Look on the Internet and, thirty seconds later, you find a design you just need to tweak.

In Chapter 1 we introduced three information-processing challenges that limit our ability to benefit from Business. First, in Gathering Information, we have vast oceans of information available about Business and our Economy. Through D-Tech, we can find much specific information rapidly. Yet, particularly in areas where we are not specialists, finding meaningful informa-tion is frequently challenging.

Second, we put that gathered information together in Understanding Context, to make sense of how our world cur-rently operates. D-Tech can both provide us great insight—and mislead us. And third, when we are Analyzing Options we must decide what path to take, knowing that many decisions are inter-connected, like pieces in a puzzle. Only if we select choices that are aligned with one another will our puzzle fit together well.

Over the coming pages, we will review how each of these challenges is relevant for our Business and Economy today. In each area, the information problems existed before D-Tech. However, D-Tech transformed and magnified these challenges. Such challenges are now driving increasingly negative results for society, while many results for Business appear robustly positive. Why? We are now asking Business to solve problems inherently different from ones successfully solved in the past.

Figure 5.2: We are each designing our own pieces
of the puzzle–and only if those pieces align will we
obtain the outcome for which we're hoping.

BUSINESS GATHERING OF INFORMATION

Consider the challenges Businesses and their customers face when Gathering Information today. On the one hand, we can gather amazing information. Want to know if Amazon offers the bicycle in front of you less expensively? Pull out your phone and do a quick search. Want to know if someone beat you to producing your improvement on the common mousetrap? Same answer.

Yet, finding what we need is not always so straightforward. We face a swarming sea of data created by our 2 billion computers. In our daily life, or where we are not deeply trained, we frequently struggle to gather meaningful information. Consider the shift in what *eating healthfully* meant for my parents versus me:

My parents wanted to eat healthfully. Thus, they ate a low-fat diet. They also ate lots of fruits, vegetables, and grains. They ate well.

Yet, I feel just a little superior to my parents. You see, I now know that fats are actually good for us (Harvard Medical School 2017).

Of course, I also realize that it's not that simple. Trans-fatty acids are a type of unhealthy fat.

Then again, I was doing some reading on the internet recently and learned that there's an exception to the exception. There are some molecules that are called trans-fatty acids that may actually be good for us (Thompson, Dennis 2015).

And then there are the saturated fats. Today, nutritionists believe that saturated fats may actually be the healthiest of all the fats for us (Ravichandran, Grandl, and Ristow 2017). On the other hand, after a 50-year battle against saturated fats, some people still believe saturated fats are unhealthful.

And then there's the mid-chain fatty acids. . .

The truth is, I don't care about any of this detail. I'm not a specialist in human nutrition. When I walk into a supermarket, I want to know which is more healthful for me: the roast beef sandwich or the chicken pot pie?

As D-Tech has become more powerful, we've obtained increasingly broad access to volumes of information. Especially in subject areas where we are specialists, that volume of information is exceptional. However, in areas we are not specialists that wealth of information overwhelms us. We rely on imprecise signals of what we want. In the example about the pie, I was looking for healthful food, but settled for the "organic" label—which only tells me how the food was grown.

Others of us are giving up: "I don't understand what's healthful now, and I never can. It's too confusing. I'm just going to buy food based on what I *do* understand: price and taste."

As we'll see, this contrast matters. Consumers, in particular, gather information about things they understand well, such as price and taste. On other dimensions, they rely on labels related to what they want, such as organically grown. Yet, as we will discuss further in a few pages, organic does *not* mean more healthful. And, when a person understands something different from what a speaker says (e.g., organic = healthful) we have problems—especially when we have the technology to change every aspect of our world.

We now turn to the second challenge.

BUSINESS UNDERSTANDING OF HOW OUR WORLD OPERATES

To fully understand the information we gather, we need to put that information into context. That is a fancy way of saying that we must use that information to infer how our world works.

Business wants their potential clients to do something: *BUY!* And D-Tech provides business the opportunity to influence clients' perceptions of how the world works—such that the clients think they *must* have the business's offering.

Before D-Tech, we had to wade through advertising—but advertising never had today's potential to influence us. The potential subtlety of such influencing can be seen in Professor Scott Galloway's (2017) experiments with Alexa. Alexa is an intelligent personal assistant that speaks with you through recently developed Digital Technologies.

Professor Galloway requests, "Alexa, buy batteries."

Amazon's Alexa replies, "Amazon's choice for batteries is Amazon Basics. AA batteries. 48 pack is $13.60 total including tax. Would you like to buy it?"

Scott replies, "No."

The automated Alexa computer software voice replies, "I also found a 20-pack of Amazon Basics AAA performance alkaline batteries. It's $7.61. Would you like to buy it?"

Again, Scott replies, "No."

Alexa seems disappointed. "That's all I can find for batteries right now."

Alexa just nudged Scott. By offering only the Amazon Basics battery brand, she subtly tried to convince him to purchase her company's battery brand. Amazon does stock other battery brands: five battery brands appear on the first page of Amazon's website under the listing for "batteries". Alexa would simply prefer you buy her company's brand. Nudging can be more blatant than the Alexa example. Just think about everything our phone knows: it sees the text as we break up with our lover, listens to us sigh as we pity ourselves, and knows what time we set up dinner appointments.

Figure 5.3: Amazon's Alexa nudges us to purchase products targeted by Amazon.

Business is looking to influence other Business as well as consumers. Business is beginning to learn to use Digital Technology and data such as LinkedIn résumés to profile companies and develop predictive sales forecasts (Kudyba and Davenport 2018). Businesses are identifying which specific buyers are easiest to influence—and whether that buyer is more likely to respond to email, phone, or some other approach.[2] At the same time, multiple companies are implementing coaching to help emails have greater influence (Finley 2016).

D-Tech brings a long journey to its climax. Advertisers began developing their craft long ago, but they have never had the opportunity D-Tech provides. D-Tech has all of the information about us. It's always with us, and we even anthropomorphize it: we talk about D-Tech as if it's human. We're used to listening to our Digital Technology. And, D-Tech can be programmed to leverage every influencing skill known to humanity, focusing most heavily on those that influence *us* most strongly.

We don't know how powerfully we can be influenced. We do know that during World War II, fascists convinced people across Europe to participate in slaughtering millions of people in gas chambers and death camps. Japanese nationalists convinced pilots to go on suicide missions.

Yet, those examples were humans influencing humans. We're still unsure of the limits of digital influence. We do know that we are convinced today to use Facebook even though the more time we spend the more depressed we become. We also do know that we tell computers the dark secrets that we won't tell a human: researchers found that medical patients were more likely to divulge the true origins of their medical conditions to a computer than to a human (Lucas et al. 2014). Does this difference in how we treat machines increase or decrease the potential for influence?

We'll find out.

We conclude this section with the third challenge: Analyzing Options

BUSINESS ANALYSIS OF OUR OPTIONS

Once participants in our Economy have gathered information and put it into context, we must decide what to do. Such decision is based on analyzing how particular actions will impact the world around us. Through D-Tech, we have amazing calculating power to analyze almost any decision. Yet, in our complex, interconnected world, that is not enough; too frequently our decisions still lead to unintended consequences.

In many ways, our D-Tech Business environment is like an accident-prone highway. In 2016, a video became popular on the Internet:

- A driver 300 yards ahead taps lightly on the brakes.
- The car directly behind brakes harder to avoid the first car.
- The third car skids across the lane and brakes harder yet.
- The fourth car brakes and swerves into the next lane over.
- The driver in the next lane didn't have time to brake or swerve...*Bam!* The front passenger side of a grey Range Rover caves in.

These drivers' decisions are interconnected: as soon as the first car braked, all of the cars needed to brake or steer. Each driver had to anticipate the decisions of the nearby drivers. As soon as one driver made a mistake, multiple drivers ended up in the accident. Perhaps most tellingly, the first driver never intended to cause vehicles several hundred meters behind him to crash. The first driver's actions simply had an unintended consequence—a *side effect*.

Our D-Tech environment is rapidly changing, like the traffic on a freeway. However, there are no lane markers. As technology progresses, we're facing increasingly dizzying arrays of options. Perhaps we're flying on a hoverbike rather than driving a car and at any given moment, we have to be concerned with vehicles above and below us, as well as in front and behind.

Figure 5.4: In our interconnected world, one person's decisions have unintended consequences on others.

Our decisions are interconnected, and *side effects* (emergent impacts) are all over the place. What type of interconnections do we have in our complex Business system? Inter-working parts, information technology (D-Tech), people, products, and more.

Through D-Tech, we've created a world with mind-numbingly complex interrelated parts and production mechanisms. The Model T might seem like it was a complex product: it used 1,481 parts (Bailey 1956). In contrast, using D-Tech, we've designed today's cars that require 30,000 parts—more than 20 times as many (Toyota, n.d.).

If I decide to change the design of any one part in the car I produce, a dozen engineering teams around the world will need to be making corresponding changes in their parts for that car. Hundreds of engineers could need to re-prioritize their days. In any integrated product or service, constructive results arise only when the parts match each other. Any decision to change any one part is likely to require aligning decisions about dozens of others.

D-Tech has also allowed us to massively increase the complexity of our production facilities and back-office support. Gerald's pen factory exemplifies such complexity. A couple of years ago, I told Gerald, "Customer Delphi wants a new type of ink. They want a blend of green #4 and red #8 dyes oxygenated for stability and delivered in the 8" hard-plastic tubes. Can we make that?"[3]

I expected Gerald would know. He had been responsible for the factory the past five years. Instead, Gerald responded, "I don't know. You've been driving us crazy doing new project after new project to add pipes between tanks, new ways to modify the ink, and dozens of unique customer projects."

Gerald continued, "The only way we can possibly control this complexity is through the plant's software. Unfortunately, as you've rushed us through these dozens of projects, we've linked all kinds of decisions in that software control system. I didn't realize the unload valves in tanks 87 and tank 124 automatically opened together until the other day. The crew meant to empty tank 87, and everyone ended up with blue ink all over their boots when tank 124 emptied at the same time."

Gerald concluded, "Physically, the plant can satisfy Delphi's request. But, I don't know if we understand how to control our own plant well enough to do so."

It's not only a plant's IT system. Every central IT control system is complex and interconnected with everything else. What's "fast" for merging IT systems after an acquisition? Two years! That's (at least) how long it will take a recently merged United Airlines to schedule all flight attendants using one system, not two. It took American Airlines more than three years when it merged with USAir (Matyszczyk 2014). It takes so long to connect such planning systems because they are connected to *everything*. Want to change how you measure overtime for flights to Milwaukee, Wisconsin? Five other systems will need to change also.

With D-Tech, our human systems are equally complex and interrelated. Consider our decision to hire technical support specialist Phil. In the 1950s, Phil might have specialized in delivering technical assistance to any ink customer and could have solved most issues. Today, Phil's the specialist in ink biochemistry; another specialist is needed when the issue is with the paper with which the ink is used. Phil knows a *lot* about ink biochemistry. However, he knows little about the impact of paper selection or the physics of pens. Phil's predecessor would have responsibility for all of those areas—reducing the number of people connected to many decisions.

Like individual parts and workers, decisions about individual products are also inter-connected with decisions about other products. When General Mills introduced "Frosted Cheerios" for the first time, for example, competition across *all* of their products changed (Hausman and Leonard 2002). We consumers looked at the healthfulness of traditional Cheerios differently; we made different decisions about what to buy. General Mills' competitor Kellogg's could now compete with traditional Cheerios using different advertising and package design.

Since we've applied D-Tech, what's happened to the number of products a Business offers? We have 275 different car models to choose from in the U.S. (Statista 2017b) and 300 types of toothpaste. Most of these products are produced by a handful of manufacturers.

Business's decisions are interconnected with Government and Cultural decisions as well. Now more than ever, Regulatory Affairs must be aligned with production. Business must identify and ensure it satisfies 250 million restrictions (across all products)[4] and a multitude of patents. (E.g., there are 486,000 patents just for cars ["Industrial Innovation" 2017]). Regulatory Affairs is more likely than ever to find surprises that require "immediate attention." Such immediate attention is likely to require re-designing one part, which, of course, requires re-designing dozens of others.

Interconnected decision-making existed before D-Tech—but never like this. As seen in Figure 5.5, today, we have one complex legacy system interconnected with another. As a result, we need to understand all the connected decisions—and then coordinate with everyone else—or respond to their mistakes. As discussed earlier in Chapter 2, this complexity is cemented into legacy systems built up piece–meal to solve specific problems as they emerged, and we need to respond rapidly because our world is changing fast. These conditions stress the limits of both humans and machines. In short, any decision in today's D-Tech age is inter-connected with dozens, if not hundreds of other decisions.

Figure 5.5: Business consists of one complex system interconnected with another complex and another. The degree of collaboration needed to change anything becomes mind numbing.

A business strategist would say we have a complex system. If everything is aligned across the system then the system can work appropriately. The challenge is in finding the appropriate process for aligning all elements.

An economist might say we have *actively dynamic externalities*. An *externality* is the impact of something you do that affects me. For example, the pain I feel from your pollution is an externality. An *actively dynamic externality* extends the concept of interconnected decisions. The concept represents he impact on me of something you do, which also depends on what Susie and I do. For example, you may tap the brakes of your car, which is two vehicles ahead of me on the freeway. The impact on me depends on how Susie, driving the car directly ahead of me, reacts. How do we coordinate in such a world where 50 or even 50,000 people might all be trying to coordinate such decision-making? That's tough.

SUMMARY

In short, D-Tech challenges Business. Across our economy, we are challenged in Gather Information, especially in areas where we are not specialists. We are overwhelmed by the volume, detail, and level of technical content. We find data, yet, too frequently, we do not find the answers to our questions. Instead, one or two word labels increasingly drive our actions—despite those labels not precisely communicating our desires.

Once we find that information, we are challenged to gain insight from it about how our world operates (Understanding Context). Through D-Tech, advertising is becoming super-powered; company's ability to manipulate our perception of the world is far greater than ever before.

Based on our understanding of the world, we set about Analyzing Options. While D-Tech strengthens ability to analyze any one decision, that is not enough. Collaboration across greater and greater numbers of people is required for success. Too frequently, instead, we find unintended consequences.

HOW HAVE OUR BUSINESSES ADAPTED TO OUR D-TECH ENVIRONMENT?

Our Businesses continue to do what they've always done: compete, cooperate, and conquer. However, each type of activity Business uses to accomplish these tasks is compromised in a critical way:

- **Product and Service Selection:** As D-Tech progresses, businesses are increasingly choosing to produce products that fit poorly with what society needs and values. Those products fit the labels for which consumers are searching (such as "organic"), but frequently miss the underlying desire (nutritious and healthful) that led to the search for the label. Where consumers have given up trying to understand a particular product dimension (e.g., nutrition), Business doesn't work on that product dimension. Thus, as D-Tech allows us to shape our world more profoundly, we are increasingly creating products with attractive labels—whose superficial beauty hides a sickly interior.

- **Targeting and Influencing Customers:** Business is just beginning to face the challenging question of, "When and how is it appropriate to influence customers' perception of the world?" We have yet to see every boardroom struggle with this question. Undoubtedly, many decisions will be made to influence customers in constructive directions.

Unfortunately, Business's record of influencing us through advertising suggests many decisions harmful to society are also likely to be made.

- **Execution:** Deep collaboration is becoming a central Business challenge. Constructive decisions must align across increasing numbers of people, machines, and systems collaborate effectively. The results of ineffective coordination are growing: destructive, unintended consequences; failed collaboration; and major issues "hiding in plain sight."

PRODUCT SELECTION

What's happening with Business product selection? In a broadening set of increasingly important areas, Business's product selection is failing society. . . leaving us sick. D-Tech *side effects* (emergent impacts) are negatively impacting the alignment between the products Businesses choose to sell and society's needs.

Earlier in this chapter, we described a D-Tech *side effect:* consumers are increasingly dependent on product labels, or giving up trying to understand particular product dimensions. Frequently, we fail to find the information we need to answer basic questions. Is an organic apple pie cooked in coconut oil healthful? I don't know; the internet has too much conflicting information to help. Some people simply give up searching. Others grasp onto labels that are available, even if they don't fit precisely: organic describes a growing regimen rather than a product's healthfulness. Yet, consumers desiring healthful food gravitate towards that label, or other aspects of an attractive (e.g., simple) label.

So, what does Business do? It produces products and services that its customers will buy. These products fit the labels such as organic—or completely ignore health concerns altogether, for example. Certainly cheap and tasty foods are popular. Meanwhile, as D-Tech allows us to increasingly customize every aspect of our products and services, this focus on labels is a growing problem. Any benefit that we can't communicate to consumers through a label is at risk of disappearing or at least being seriously weakened.

"Organic" may be an attractive label. In fact, organic food now accounts for more than 5% of all food purchased in the U.S. (Organic Trade Association, n.d.) and we pay almost 50% extra for these foods versus others (Consumer Reports 2015). Organic foods may even be modestly more healthful (Brandt 2012). However, the term organic does not *mean* it is automatically healthful. A nutritionist would likely suggest a (well-washed) traditionally grown red apple (11-15% sugar) is more healthful than an organic Fuji apple (16-18% sugar).

Low Fat? Likely unhealthful. Much of the fat is likely to be replaced with sugar to maintain the flavor (Wansink and Chandon 2006).

How about more nuanced aspects of food labels? Nuances don't really help. Food manufacturers regularly invest in making their labels appear more healthful. For example:

- Reducing the number of ingredients to make the label appear "clean."
- Replacing ingredients whose names are complex.

Unfortunately, such actions frequently lead to less healthful rather than more healthful foods. For example:

- Additional processing may be required to retain flavor.
- The ingredient with the more healthful-sounding name may actually be less healthful.

Lack of focus on what the client truly desires (*healthfulness*) further damages our products, particularly food, indirectly. Customers focus on food's price (labeled), taste (easily recognized), and convenience (easily recognized). As a result, much of food science focuses on "size, growth rate, and disease resistance" (Scheer and Moss, n.d.). Rapid growth and disease resistance reduces cost. More convenient size makes the food easier to put into our kids' lunches. We get the characteristics on which we focus: size, cost, and convenience. We like that. We buy that.

Unfortunately, the result is sickening—literally. Rapidly-growing, disease-resistant foods tend to be less healthful:

- Fruits and vegetables have less protein, calcium, phosphorus, iron, and riboflavin than they did in 1950 (Scheer and Moss, n.d.).
- A chicken in 2004 contained more than twice as much fat as in 1940, a third more calories, and a third less protein (Purvis 2005).
- Recently bred apple varieties (including Pink Lady, Fuji, and Jazz) are bred with growing sugar content to appease our sweet tooth (Torrens 2017).
- Phytonutrients, keys to reducing cancer, cardiovascular disease, diabetes, and dementia, are decreasing. Phytonutrients tend to be lower in tasty food (Robinson 2013).
- Organic foods don't offer the healthfulness that farmers provided in the 1940s, 50s, and 60s (Plumer 2015).

Labeling problems with our physical products and services existed before D-Tech. For example, every seven or eight years we bred a new generation of cows—focused on farmers getting more and cheaper milk. There were likely *side effects* of that.

However, with D-Tech, our technology is increasingly fast-acting with broad capability. In addition to breeding plants and animals, we can modify genetics, pinpoint specific nutrients, and even chemically synthesize or concentrate particular molecular compounds. Farmers, grain handlers, food processors, food manufacturers, and grocery stores can all modify aspects of our food—targeting one product label or another. Whether or not such activities decrease healthfulness in some unlabeled manner cannot be the focus of a profit-driven business Business: with the label remaining the same, consumers are unaware of the difference, and thus cannot pay the Business for that (unlabeled) difference.

Labeling problems can be even more extreme for virtual products and services. James Brindle (2017) describes what happens to the search terms (the labels) we can use to find children's videos.

With D-Tech, a team of animators can automate mass production of children's videos. We certainly get a vast array of videos for "Finger Family Song," "Learn Colors," "Superhero," "Animal Sounds," and "Excavator."

Unfortunately, such automated production is low-quality. Many children's videos on the internet are "a nightmarish mishmash of cheap-looking automated 3D animations created from character models and motion-capture libraries." Other videos are humans acting out weird scenarios to fulfill titles a search engine is likely to find, such as "Batman Finger Family Song—Superheroes and Villains! Batman, Joker, Riddler, Catwoman." "There is, in fact, no editorial control. Just a mathematical probability expressed via an algorithm being acted out and auto-played over and over (Major 2017)."

Today, people watch an estimated five billion YouTube videos a day (Danny 2018). A good number of those videos are high quality. Parents that take special care in searching through the videos *can* find high-quality entertainment for their children.

These parents can also find specially curated videos at Netflix or other sites if they use just a little bit more money, time, and care.

Yet, with D-Tech, the relative ease of finding low-quality and high-quality children's entertainment has flipped. Too many times, I've found myself letting my kids watch some low-quality video found on YouTube. I rationalize by thinking, "Well, that video may help them learn to count." Nonetheless, the video truly is turning my kids' brains into mush. Product selection is increasingly problematic in our D-Tech world.

Such negativity contrasts with the direct, amazing potential of D-Tech such as an electronic chip that reinforces human memory to mitigate short-term, memory loss (Kurzweil 2018). Another amazing product example is new headgear by which we can control prosthetic limbs with our thoughts—and then *feel* the objects with which the prosthetic limb interacts (Prattichizzo et al. 2018). Those are real—and amazing—products that D-Tech has helped us create.

Dream your most amazing dream—and then type that dream product or service into a search engine's tiny little box. Then press "search." Frequently, we can obtain exactly what we want, and D-Tech has helped make those products and services *amazing*.

However, there are destructive elements to our current D-Tech environment. We're confusing ourselves and giving up where we can't afford to. Any benefit that we cannot clearly label is at risk of disappearing or at least being seriously degraded. These negative impacts are D-Tech *side effects*.

How about the way Business executes its activities once it knows which products and services to deliver? How are D-Tech *side effects* impacting these?

TARGETING AND INFLUENCING

Much of Business is still deciding how it wants to unleash D-Tech's most-powerful, influencing capabilities. Unfortunately, Business has a record of influencing us with advertising that is more destructive than constructive.

Consider Business's history in advertising food to children. A clear and unattractive picture emerges. What foods are advertised to children (Federal Trade Commission 2012)?

- Added-sugar drinks
- Traditional U.S. snacks such as chips
- Candy and frozen desserts
- Fast food ("Quick Service Restaurants")
- Low-sugar cereals [5]

Less than 0.5% of food advertised to children is currently focused on fruits and vegetables. Presumably, the "voluntary initiative to advertise healthier options to kids" rewards society with that 0.5% of healthy advertising (Barclay 2016). Food advertised to adults follows a broadly similar pattern (Statista 2017b).

Many fruits and vegetables are unbranded and hence challenging to advertise profitably. Nonetheless, this challenge seems more of an excuse than an explanation for why *almost nothing* healthful is advertised.[6]

Such negative outcomes are consistent with our generally negative analysis of the impact of D-Tech on American Culture discussed in Chapter 3. Such precedents are unnerving provided the potential damage of D-Tech.

The article, "Untapping the Unconscious: The Next Revolution in Pharma Marketing" (Sharman and Allman 2013), is freely available on the internet. The article highlights how physicians in Ohio have "nudged" their patients to increase regular cancer screenings by adjusting default settings in electronic forms to

"Please automatically sign me up for cancer screenings" rather than a default of "Do not sign me up". The article also highlights that, "all choice is influenced. There is no such thing as neutral choice architecture. So, why not nudge people?" Geyer (2018) suggests pharmaceutical marketing companies "strive to deliver personalized content via [digital] profile (not health status) and digital interaction preferences."

Perhaps my concern is too strong. Undoubtedly, there will be socially constructive influencing. Medtronic and IBM, for example, are developing new methods to help diabetics manage their disease. They inspire us with tales of the problems they will fix: "Every day, I think about a 15-year-old girl who we couldn't save because her sugar levels got too low and she was sleeping and couldn't wake up to raise them on time (Medtronic, n.d.)." Watch manufacturers convince us they'll keep us exercising (Bonnington 2014). And there are at least 10 manufacturers who claim they'll improve our focus and attention (Glei, n.d.). Meanwhile, even the articles begging pharma manufacturers to nudge their potential clients more aggressively include suggestions that the business does so, "toward better patient outcomes (Sharman and Allman 2013)," and by providing "personal patient videos to make explanation more accessible and comprehensible (Geyer 2018)."

Initiatives to ensure constructive influencing are also being developed. The IEEE (The Institute of Electrical and Electronics Engineers) is in the early phase of developing standards for "Ethically Driven Nudging for Robotic, Intelligent, and Autonomous Systems" (IEEE, n.d. "IEEE Project 7008"). The organization, Time Well Spent (timewellspent.io), is "dedicated to reversing the digital attention crisis by realigning technology with humanity's best interest." Time Well Spent is working to Inspire Humane Design, apply Political Pressure, inspire a

Cultural Awakening, and Engage Employees. Such standards and initiatives are positive. Perhaps such initiatives will change our world.

In short, each business today decides how to focus its societal influence. There are projects we can categorize as positive; others as negative. There are also impressive initiatives to potentially curb some of the most negative options. However, our experience with traditional advertising-based influence suggests the negative impacts will win out unless we change direction.

EXECUTION

Our businesses are behemoths. Their leaders are now challenged to interpret complex amounts of data, interconnect decisions, and coordinate hordes of increasingly specialized humans. If today's organizations were machines, an engineer would say they are "challenging to control."

Changes within an organization have become rabid dogs chasing an organization in circles. Suppose you hire Phil without aligning with your colleagues first. That change barks at the sales director. The sales director is going to feel the pressure to consider new client targets. As sales successfully woos these clients, this change barks at Susie. Woof! More and more salespeople begin to pressure Susie to change her product mix. And then Gerald gets bitten on the ankle. He's asked to contribute more and more specialty engineering support of a type he doesn't have on his team.

Or, changes become rabid dogs fighting each other. You hire Phil to create innovative, new products with customers, which tend to complicate production facilities. At the same time, Gerald hires an engineer specializing in simplifying production facilities. Those two conflicting decisions may meet in a dogfight shortly.

What's happening? Our problems are interconnected.

Consider three related challenges Business is facing today: rising unintended consequences, failed collaboration, and major issues "hiding in plain sight."

We face a march of unintended consequences. In the light of today's complexity, too frequently we have preconceived (and false) expectations for what will happen. Consider one recent attempt to improve American healthcare.

Society's leaders thought they knew what to do. The poor obtain a lot of (expensive) emergency room care. These leaders thought the poor used emergency rooms only because they didn't have access to traditional care. These leaders thought, "Give more poor people health insurance. If we do that, the poor will get traditional doctor care instead of using expensive emergency rooms." The result of this thinking was featured in the Affordable Care Act (ACA), which gave many poor people health insurance.

Wrong move. Emergency room use by the newly insured poor went *up* 40% to 65% with the ACA, not down (Taubman et al. 2014). The poor people using the emergency rooms were frequently extremely sick people. These people did use their new insurance to get traditional doctor care. However, this regular care didn't materially reduce emergency room use. In fact, traditional doctor visits increased emergency room use: the doctors found additional conditions needing treatment.[7] Perhaps giving poor people health insurance makes society more fair. However, it doesn't reduce emergency room use. The ACA designers misunderstood the side effects such insurance would have on healthcare. Disruptions, industry responses that seem to "come from nowhere," and unmanageable situations exist across Business today.

Bill was spending days-on-end lying stiff in his bed, grimacing from excruciating pain. He eventually agreed with Dr. Corette that he needed a surgically implanted pain pump. However, Dr. Corette believed Bill's insurance company required a test with the appropriate pharmaceutical product before paying for the pain pump and related surgery.

Dr. Green's office appreciated Dr. Corette's referral and the prescription for "pain medication application under medical guidance." However, Bill's medical file was dozens of pages thick. Dr. Corette skimmed through the top few pages of the file until he felt comfortable he knew what to do: Bill had multiple sclerosis and needed pain medication. That sounded fairly straightforward.

Unfortunately, Dr. Green received none of the message about "testing medication for a surgically-implanted pain pump," and "insurance company as the primary target for the information."

What happened? Bill did receive a pain medication application. However, this was the unintended outcome for Bill:

- Received a specialized medication application that caused his body to respond differently than it would if the medication had come from an implanted device. As a result, Bill didn't learn which medication a new surgically-implanted pain pump should deliver to him. Dr. Green's office hadn't realized the purpose of the test.

- Received a $2,000 bill: Dr. Green's office had not focused on the insurance company because it had not been alerted to the importance of that issue.

Figure 5.6: Simultaneous efforts to streamline
for a simple factory and to add value with more
complex products can meet in a dogfight

Even when we know what we want to do, implementing our
plans can be more challenging than we expect. What happens
too frequently when different specialists need to collaborate on
patient treatment?

The story about Bill is true. Unfortunately, that story is far from unique. Spend time with any of the millions of Americans receiving care from multiple physicians, and you'll likely hear a similar nightmare. In fact, spend time at the water cooler of many American companies and you'll hear similar stories about lack of coordination among different businesses, functions, or key individuals. We finally understand what we're going to do, we finally figured out how to pay for it—and we still find a way to botch it—because our actions need to be interconnected with dozens of others.

Of course, those challenges thus far relate to challenges we decide to solve. We also have too many problems that, from the outside, appear like no one is trying to solve. Today 28% of Americans age 40+ are prescribed statins (Salami et al. 2017). Unfortunately, as currently prescribed, these pharmaceuticals appear to have a negative impact on health (Topol 2013).

Outside healthcare, the solution seems straightforward: stop prescribing statins, or, at least, prescribe them less frequently. Within healthcare, the problem is more challenging. Coronary heart disease, which statins attempt to mitigate, is a national problem. Statins are an effective solution to some problems. Moreover, no one within our healthcare system "owns" the responsibility to guide constructive progress on the use of statins: patients, hospitals, insurance companies, regulators, and physicians all have a role to play.

The human-made element of such problems is more challenging than those faced by Horacio and Daksha during the Agricultural Revolution thousands of years ago. Perhaps Horacio and neighbor Luize both wanted to use a particular hoe owned by a third neighbor, Ignacio. Horacio and Luize debated and argued for several hours before Ignacio simply asked, "To borrow my hoe, who will pay the most?"

For Horacio and Daksha, such coordination with neighbors was certainly challenging. Today, we ask questions that seem, at least to us, more complex, such as, "How should we refer a complex medical patient from one physician's office to the next?"

Business is certainly taking actions to address complexity. Spend a few days in a business and you will see D-Tech control systems everywhere: spreadsheets to manage customer receivables, automated systems to control the pumps and valves in the factory, Enterprise Resource Planning systems to manage everything from inventory to sales. D-Tech is critical for managing the complexity in today's businesses. Deep collaboration among many individuals is challenging to achieve, no matter the technology.

GOVERNMENT AND CULTURAL INFLUENCES

Our Government and Culture have also increased Business's challenges. In Chapter 4 we discussed how Government is littering our environment with excessive yet incomplete regulations. Here, we simply highlight that those failures have had negative impacts on Business. If one element of society has a severe breakdown, it negatively impacts other aspects. It's like a car with a torn fan belt: if there is only a small tear, other parts of the engine can compensate. However, once completely broken, a wildly flapping fan belt can smash other engine parts (Figure 5.7).

We also discussed that D-Tech has played a critical role in the U.S. Government's complexity. After close to 50 years of applying D-Tech, the U.S. Federal Government has grown to 2.8 million employees across 250 agencies enforcing over 250 million restrictions in total. Every business must comply with each of these hundreds of millions of regulatory requirements—with compliance costs potentially consuming as much as 25% of our economy. Despite such efforts, as we discussed in the last chapter, the average professional *accidentally* commits three felonies a day (Silverglate 2011).

Figure 5.7: A single broken element in a complex system can demolish all of the other pieces in that system.

Business must evolve in response to such D-Tech *side effects*.

In Chapter 3, we discussed D-Tech *side effects* on our Culture. As we increasingly apply D-Tech, we are becoming less engaged in local society. With D-Tech communication, we are less and less limited to communicating only with people physically close to us. We can still walk out our front door and speak with a neighbor; yet, increasingly, we choose instead to invest more of our social capital in developing relationships with those physically farther from us.

Unfortunately, such relationship investment patterns leave us feeling alone. Our close friendships suffer. We miss the ongoing, extensive friendships and family connections that used to be within our reach. We'll return to this topic when we discuss healthcare and education.

WHAT IS THE IMPACT OF BUSINESS ADAPTATION?

Business is failing to bringt the tidal wave of D-Tech's potential to society.

As we discussed in the last section, the *side effects* of D-Tech are changing the nature of the economic problem. Increasingly central for Business today are problems of coordination and relative access to information. On the horizon are challenges of D-Tech's ability to influence us. No computer is actively destroying a company's data sets. No cell phone is scrambling our voices. Nonetheless, the breakdowns in Business are *side effects* of our applying D-Tech:

- D-Tech's capabilities are impressive: unparalleled access to information and people, exceptional ability to influence our understanding of the world, and unrivaled capacity to analyze options.

- As we apply these capabilities, we are changing our environment, creating a virtual world that changes the nature of the most vexing problems Business faces:
  - Finding information and answers in areas where we are not specialists.
  - Becoming exposed to increasing influence, especially in areas where we are not specialists.
  - Collaborating across specialists.
- As we discuss in this section, Business is breaking down as a result:
  - Our immediate challenges are severe. Today, D-Tech *side effects* leave more than 60% of the U.S. economy failing to deliver what society needs.
  - Tomorrow we face potentially even greater challenges— we are becoming increasingly dependent on the decisions of a few Business leaders.

IMMEDIATE CHALLENGES

What is happening? D-Tech has changed the heart of the economic problem that Business is trying to solve. Increasingly important are challenges in collaborating and finding answers. Breakdowns in our Culture and Government are magnifying these challenges. On the horizon are additional challenges originating from D-Tech's growing ability to influence us.

Business is failing to solve such problems in society's best interest. D-Tech *side effects* leave more than 60% of the U.S. economy dedicated to industries failing to deliver the basics society needs. Also, business is not delivering the economic expansion D-Tech promises.

Figure 5.8: Even the best-looking clay snakes don't satisfy the need for plates; today, in too many circumstances the U.S. economy is producing clay snakes when the need is for plates.

The problems that were primary before D-Tech remain important. As an economist might highlight, resource allocation remains critical. The competitive, market-based system (i.e., capitalism) must remain central to Business's operation.

Nonetheless, our new D-Tech *side effects* are increasingly problematic. Examples of the resulting inefficiency and destructive outcomes abound in our next section.

FAILURE TO DELIVER WHAT SOCIETY NEEDS

<u>The majority of the U.S. economy is dedicated to industries failing to deliver what society needs</u>. Suppose a worker named Rayden lives in a society that desperately needs as many plates to eat from as possible. However, Rayden isn't actually making plates. He's making little squiggly snakes with clay. The snakes are fun. However, society doesn't need clay snakes—we need plates! This problem is indicative of what we're facing today.

In our real world, we're not talking about plates. We're talking about Business's increasing failure to deliver our most fundamental requirements. A healthcare education tool that augments reality is impressive (Sweifach 2017). Our fundamental need, on the other hand, is to be healthy and alive. We're not. American life expectancy fell over the past two years (Economist 2018).

The failures run far more broadly than healthcare. The majority of the American economy (across Business and public services) is similarly failing. Business is not delivering our most basic needs.

In this next section, we discuss these fundamental failures market by market—as well as the role D-Tech *side effects* have had in those failures. Figure 5.9 shows the markets we discuss as fundamentally failing to deliver what society needs: more than 60% of the U.S. economy. We also discuss the U.S. labor market's failure to deliver our most basic needs in that market.

Each failure is unique. As a result, this next section is long. As a reader, use the information you find valuable. Two possible approaches:

1. Focus on the market of greatest relevance to you.
2. Begin with the financial market, which has the most direct connection between the market failures and D-Tech. Follow

with other markets where the outcomes are more indirectly related to D-Tech.

U.S. Markets by Share of Economy

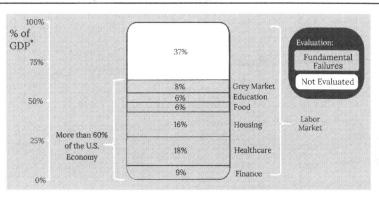

More than 60% of the U.S. economy is dedicated to markets failing to deliver what society needs.

*#s from sources referenced in the following, should solely be considered indicative.

Figure 5.9: This chart illustrates that more than 60% of the U.S. economy is dedicated to markets fundamentally failing.

At the end of each of these discussions, we've put links to help you navigate to the next section you'd like to cover.

FINANCIAL MARKET—9% OF U.S. ECONOMY (GDP)

Our financial markets are 9% of the U.S. Economy, by GDP (Greenwood and Scharfstein 2013). These markets are failing. The purpose of our financial markets is to transfer funds efficiently while providing quality information.

The financial markets transfer funds from people with savings to people wanting to invest. For example, suppose Primod wants funding to build his robot toy. He starts a GoFundMe campaign. The GoFundMe software, operated by a business in the financial industry, sends those of us who want to invest in the robot's development an email about the campaign. Savers and investors are brought together, and the financial market benefits from the fees it charges for the service.

In a well-functioning financial market, the prices of instruments traded by the financial markets also play a critical role in society. Those prices provide information about relative scarcity. Consider, for example, futures prices for corn and sorghum. Farmer Lize bases her decision on which grain to plant for next year based on those prices. If the price of corn is high relative to sorghum, she plants more corn. When the price reflects the relative scarcity of corn and sorghum, then what Lize decides is efficient for society: a high price of corn reflects that corn is scarce, and society benefits when Lize plants more of it. Competitive forces across our economy reinforce a healthy relationship between price and relative scarcity.

But not D-Tech. As illustrated in Figure 5.10, D-Tech *side effects* are causing financial prices to diverge from relative scarcity. D-Tech's rapid, automated, trading mechanisms have interconnected trading decisions, seemingly leaving financial market prices a less accurate reflection of relative scarcity. Too frequently, terms like "bubbles," "excess volatility," and "information cascades" are associated with our financial market prices today.

The following case of Navinder Sarao illustrates the relationship between D-Tech and financial market price instability:

Navinder Sarao was unshaven and sported a bright yellow shirt as the judge intoned, "Mr. Sarao, you are accused of one count of wire fraud, ten counts of commodities fraud, and one count of 'spoofing'."

Mr. Sarao winced. He was just charged with causing financial prices to crash in an event called the Flash Crash of 2010 (Levine 2015).

According to the legal complaint, Mr. Sarao fraudulently "spoofed" orders to sell $200 million in stocks. Programming his computer to send orders to sell $200 million in stocks—but then cancel the orders instantly before those orders were executed— was the core of the alleged spoofing.

On the fateful day, Mr. Sarao's big lineup of $200 million in sales appears to have caused a cascade of events. Many viewing Mr. Sarao's sell orders recognized that so many sales would cause stock prices to dip—and these observers tried to sell their stocks before the dip. Yet, those rushed sales also caused prices to fall. D-Tech software recognized that prices were falling, and the automated trading software piled in—everyone trying to sell before everyone else. Prices fell faster and faster, until a full $1 trillion in stock value had disappeared (Cassidy 2015).

D-Tech's automated, incredibly rapid traders are key actors driving such price shifts. Such automated trading can cause small price shifts to become large ones. Academics reference theories of information cascades and price bubbles (Brunnermeier and Oehmke 2012).

It's as if financial market decisions are a line of dominoes. Before D-Tech, each domino was generally separate from the other dominoes. Today, D-Tech decision-making is lining up dominoes only millimeters apart—as soon as one domino falls, another follows immediately.

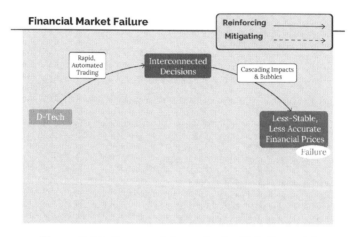

Figure 5.10: Current failures in the U.S. financial markets are side effects of D-Tech: through D-tech, increasing numbers of trading decisions are becoming interconnected, which is leading to pricing failures.

Professor Frank Zhang of Yale University statistically confirmed the more machines that trade stocks, the more volatile financial market prices becomes (Zhang 2010). In addition to the Flash Crash of 2010, we've seen high-profile recent cascade-driven crashes in the cryptocurrency Ethereum (Watts 2017) and the stock for Amazon.com (Durden 2017c). In fact, in August 2015, the Nasdaq stock market reported 1,200 halts in trading to stop panic crashes (Egan 2015).

We've similarly experienced financial bubbles in crypto-currencies such as Bitcoin: "People are comparing Bitcoin to tulip bulbs. I think those comparisons are apt. But at least with tulips, you had something tangible—a plant (Dillian 2017)." Bubbles in the value of startups: The car company Tesla, for example, is worth more than Ford plus General Motors despite having only 1% of those companies' sales (Molla 2017). Ride sharing (taxi) service Uber has few assets, yet it is worth more than Ford plus Honda, as well as more than General Motors (La Monica 2015). And home-sharing platform Airbnb owns no hotels, yet is valued as highly as Hilton and Hyatt hotel companies—combined.

Finance Professor Didier Sornette's ETH Zurich website maintains an engaging quantitative record of such financial bubbles ("Financial Crisis Observatory," n.d.). What information does such a record imply? As the respected financial analyst Mauldin Economics describes, "In 2017, we had the 'everything bubble' (Dillian 2017)."

Like dogs chasing their own tails, digital, automated trading machines seem to get excited when they see any small price move. "The market's moving up!" thinks the machine. "I gotta keep chasing that price up!" and the market price keeps going up as a result—Zip . . . UP! Zip . . . DOWN!—with the automated trading machines chasing frantically behind.

Farmers making decisions on how to plant their crops are no longer basing their decisions on the supply and demand for their crops; they are basing their decisions on robotic dogs chasing their own tails.

Not everything is going wrong in our financial markets. Our financial markets now offer a spectacular assortment of ways to invest our savings or get funding for your business. Moreover, in the long run, and even medium term, investment prices do, arguably, represent scarcity.

In addition, rapid, crashing prices have occurred in financial markets at least as far back as 1623 ("List of Stock Market

Crashes" 2018). The Flash Crash of 2010 may not have been Mr. Sarao's fault. Other traders or other factors could have been the primary trigger.

Figure 5.11: Current failures in the U.S. financial markets are side effects of D-Tech: through D-tech, increasing numbers of trading decisions are becoming interconnected, which is leading to pricing failures.

Nonetheless, D-Tech is exacerbating serious problems. Decisions in financial markets are more tightly interconnected than before D-Tech. As a result, decisions to buy or sell have more important, *indirect* impacts than ever before—with the resulting prices less aligned with real world scarcity. Price bubbles seem to be floating across our financial landscape. The bubbles continue to inflate until . . . POP!

For those following the text of this book, we now turn from our financial markets to our healthcare markets. You may also choose to read these sections out of order, or simply to skip to our discussion on <u>Risks to Our Future</u>.

HEALTHCARE MARKET—18% OF U.S. ECONOMY (GDP)

Our healthcare market, 18% of the U.S. economy (CMS 2018), is failing. As described previously, the purpose of healthcare is to efficiently keep us healthy and living a long life. Our healthcare businesses and public services are not delivering. And the D-Tech Revolution is at the heart of this failure's expansion.

Even before D-Tech, the complexity of the U.S. healthcare system was unparalleled (see, for example, Tomes [2017]). We had Medicare Parts A, B, C, and D; 50 different state Medicaid programs; Medicare Advantage; Medicare Gap plans. And that's just the start. Even before D-Tech, the U.S. healthcare system was "exponentially more complicated than the system in any other country" (Taylor and Morrison 2011).

Yet, that inefficiency only reached a critical level when we introduced D-Tech. Figure 5.8 (based on Ortiz-Ospina and Roser 2018) summarizes the failure of the U.S. healthcare system over the course of the D-Tech Revolution. This figure captures the results of healthcare markets in developed countries (OECD) over time.[8] The vertical axis shows life expectancy in years. The horizontal axis shows total money spent on healthcare per person, per year (private plus public). In other words, we spent just over $1,000 per person on healthcare in 1970 and had a 71-year life expectancy; today, we spend more than $8,000 per person and can expect to live about 78 years.

Life Expectancy vs Health Expenditure

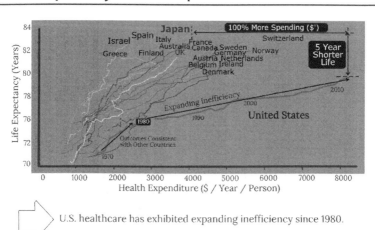

U.S. healthcare has exhibited expanding inefficiency since 1980.

Based on Ortiz-Ospina and Roser (2018) "Financing Healthcare". Our World in Data. https://ourworldindata.org/financing-healthcare.

Figure 5.12: U.S. healthcare exhibited expanding inefficiency since 1980, with today's results particularly poor relative to those in other developed countries.

The U.S. healthcare experience was consistent with the experience in other developed countries until about 1980, ten years into the D-Tech era. Until 1980, we spent more on healthcare each year and lived longer each year. This spending had a positive impact.

From 1980 on, however, our healthcare market experience changed. We began to spend much more each year, but received only a modest increase in life expectancy. Today, we spend about double what people in the average, developed country spend. Despite that spending, we have the shortest lifespan among these developed countries (Rogers 2016). In both 2016 and 2017 American life expectancy has actually fallen (Tinker 2017; Economist 2018).

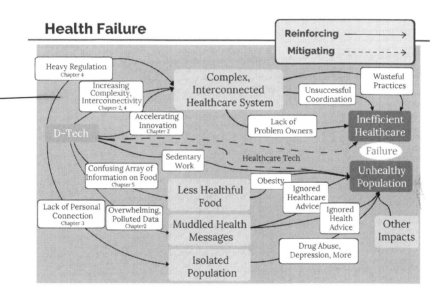

Figure 5.13: Current failures in the U.S. healthcare markets are side effects of D-Tech: through D-tech we've created an increasingly complex, confusing health environment, featuring less healthful food. The results have been poor.

The *side effects* of D-Tech on our healthcare market that led to such failures are illustrated in Figure 5.13:

- Through D-Tech, we developed a system with spectacular levels of inefficiency by multiplying the complexity of the healthcare systems.
- Through D-Tech, we triggered unhealthful behaviors by providing less healthful food, muddling health messages, and weakening human connections.

Let's begin by discussing the top portion of the figure—the origins of U.S. healthcare inefficiency.

Why has U.S. healthcare inefficiency expanded so rapidly during the D-Tech era? We have created almost unimaginable levels of complexity at every level of the U.S. healthcare industry: patient test and treatment, business system, and regulatory.

Let's begin with the complexity. As mentioned previously, through D-Tech we've come to recognize 37.2 trillion cells in the human body (Eveleth 2013). As D-Tech accelerated our ability to create new technology, we've developed thousands of new tests and treatments. Today, thanks to creativity D-Tech inspired, the average primary care physician (PCP) works with 370 unique diagnoses, 600 medications, and 150 unique lab tests (Institute of Medicine 2013, 71).

In addition to treating patients on their own, the PCP is also responsible for coordinating patients' medical care. And, as our understanding of our bodies has accelerated with Digital Technology, the number of specialists has exploded. The average PCP today is expected to coordinate care with more than 229 doctors who work in 117 patient practices (Pham et al. 2009).

The business system employing all of these physicians, tests, and treatments, has also surpassed a critical level of complexity. Suppose you go to an American emergency room (E.R.) within your healthcare insurance network (i.e., you go to an E.R. accepted by your insurance company). Your use of the E.R. room will be billed as "*in network.*" However, there's a 22% chance that the doctor who works with you inside that room will be billed as *outside* of your healthcare insurance network.

How does such E.R. billing work? The concrete building that holds the E.R. has a different relationship with your insurance company than the doctor does. Because of D-Tech, we can define any relationship puzzle piece and record its interconnection with other pieces. This permanent, automated record is a gift of D-Tech. The resulting complexity is the *side effect*.

All of these Business relationships sit within a complex Government environment. U.S. healthcare must abide by, literally, millions of constraints. Perhaps the coding requirements summarize the situation best:

There are 69,823 official ICD-10 codes (AAPC, n.d.). Healthcare workers are required to use these codes to categorize each diagnosis. The list of diagnoses is sufficiently complete to differentiate the "first" and "subsequent" times someone is sucked into a jet engine, the first and subsequent encounters when someone is struck by a duck, and the first and subsequent outings in which patients are burned when their water skis catch fire (Williams 2015). How do you create such an absurdly intricate list? Only with Digital Technology.

Such a complex healthcare system suffers the types of inefficiency described in the previous section about D-Tech *side effects*. Wasteful and harmful practices crisscross such complexity. How about that primary care physician who administers 600 medications and 150 lab tests—while also coordinating with 229 *other* doctors across 117 different practices? D-Tech may help the physicians coordinate superficially— recommending drugs less likely to have harmful side effects when combined, for example.

However, the most effective decisions each physician makes are clearly interconnected with those of the others; these decisions are all impacting the same patient's body. And, there's a good chance the patient has four or more chronic health conditions (31% of today's Medicare patients do). Could the arthritis pill make it more likely the patient eats less healthfully, which causes the diabetes to act up? Sure. Nonetheless, that type of interconnection is likely to be identified only with quality alignment. Such meaningful coordination *cannot* occur within a 24-hour day. There are simply too many interconnected decisions. The indirect impacts across these complex interactions are unlikely to be discussed and negotiated among all of the doctors working with any single patient.

Failures are hiding in plain sight.

- Today, 28% of Americans age 40+ are prescribed statins (Salami et al. 2017). Unfortunately, as currently prescribed, these pharmaceuticals appear to have a negative impact on health (Topol 2013).
- Every year, 30 million men undergo prostate exams (PSAs) at a cost of $30 billion (Mahar 2010). However, the inventor of the test describes the PSA exam result as a "coin toss" that "has not reduced the death rate in men 55 and over (Mahar 2010; see also, Skinner 2016)."
- As many as 76% of adult women receive a breast cancer screen (Health Resources and Services Administration 2017); despite the cost, all of this screening has done nothing to reduce the rate of advanced-stage breast cancer in women (Welch, Gorski, and Albertsen 2015; discussed in Belluz 2015).

Such inefficiency has gone beyond newly invented tests and treatments. More than 92 million U.S. adults (38% of the population) were prescribed opioids in 2015 despite 11.5 million people (5% of the adult population) reporting illegal use, and

1.9 million (0.8%) reporting full-fledged addiction (Han et al. 2017).

In addition, "Virtually every family in the country, research indicates, has been subject to over-testing and over-treatment in one form or another." A physician profiled eight patients visiting a clinic one afternoon. "It appeared that seven of those eight received unnecessary care. Two of the patients were given high-cost diagnostic tests of no value. One was sent for an MRI after an ultrasound and a biopsy of a neck lump proved suspicious for thyroid cancer. (An MRI does not image thyroid cancer nearly as well as the ultrasound the patient had already received.) Another patient received a new, expensive, and, in her circumstances, irrelevant type of genetic testing." And on and on (Gawande 2015).

Such a result may seem extreme. Yet, unfortunately, the fundamental cause—poor collaboration—is common. As Taylor and Morrison (2011) describe, "Many millions of people follow [healthcare] reform debates with interest and passion. However, because the issues are so complicated, very few understand them. Which is why rhetoric often trumps substance, and misinformation often fuels strong opinions. And why American healthcare is likely to be extraordinarily inefficient and expensive far into the future."

As Dr. Gary S. Kaplan (2017) summarizes, "I cannot recall a time during my 35 years in healthcare when changes in reimbursement models, government regulations, quality measures and cost controls were not being debated, planned or implemented." Nonetheless, neither the political party driving the reform agenda nor the changes implemented impacts the result. Our expanding inefficiency continues.

This D-Tech *side effect* of complexity, in other words, is a primary determinant for why our healthcare system is so inefficient.

As discussed previously, this complexity certainly leads to mistakes: D-Tech is the third-highest cause of death in the U.S.. However, for a fuller perspective on our lack of good health, we need to look at the *side effects* of D-Tech on our Culture.

As illustrated in Figure 5.13, D-Tech *side effects* also play a critical role in our becoming an unhealthy nation.

D-Tech *side effects* play an important role in our eating less healthful food, reacting poorly to messages about our health, and feeling isolated. These issues all have negative impacts on our health.

Our health depends on our own behavior—not just some outcome delivered by the healthcare industry. Chronic disease accounts for 86% of U.S. healthcare spending today (CDC 2017). Patients never fill almost one-third (31%) of the pharmaceuticals prescribed (Crawford 2014).

Disease does not simply appear from the heavens to strike unsuspecting Americans. According to the Kaiser Family Foundation (Heiman and Artiga 2015), the health of a population is explained by this descending order of importance:

- 40% by lifestyle choices (e.g., eating and sleeping habits)
- 30% by genetics
- 20% by our environment (e.g., living in carcinogen-filled areas)
- 10% by our healthcare system

How do these four factors impact our health?

Jim smokes cigars and eats hamburgers with fries six days a week. Jim doesn't exercise, is obese, and gets angry frequently. This lifestyle increases Jim's chance of a heart attack. Coronary heart disease (including "heart attacks") is the highest-cost health condition in the U.S.

Emiel drinks too much and is also obese. He eats cheap peanuts, corn, soybeans, and cheese. There's a good chance some of these foods are contaminated with a family of fungi called aflatoxins. Emiel is also exposed to a number of chemicals at work, especially benzene. This lifestyle and environment increase Emiel's chance of getting cancer. Cancer is the second highest-cost health condition in the U.S.

Paul smokes. Smoking increases the chance Paul gets chronic obstructive pulmonary disease (COPD), which makes him hack and cough. COPD is the third highest-cost health condition in the U.S.

We can do this same exercise for all of the 10 highest-cost health conditions. In total, these conditions account for more than half of total U.S. healthcare cost. We could probably do the same exercise for the top 20 highest-cost health conditions, which account for more than 80% of U.S. healthcare cost. When we make unhealthy choices, we are more likely to get sick. When we allow ourselves to stay obese, sleep poorly, and remain sedentary, our healthcare costs likely increase.

And once we are sick, how do we get better? We choose to follow the doctor's orders and take care of ourselves. Or, we choose to ignore the doctor's prescription, we return to chain smoking—and we end up back in the hospital.

Our Culture plays a critical role in determining how well we take care of ourselves. A Culture that highly respects doctors is more likely to have patients who follow the doctor's orders. In many cases, Culture (combined with financial situation) has a bigger impact on health outcomes than does the hospital where we're treated.[9]

How strong of a statement is the thesis that: "Culture is more important than hospital choice?" Very strong. A patient seeking opinions from two doctors will more than likely receive a materially different treatment recommendation from each doctor (Chow and Coye 2013). Nonetheless, socioeconomic status (highlighting Culture) is an even more important determinant of health outcome than is physician choice.

So, what are the relevant D-Tech *side effects* on our health habits and Culture?

D-Tech has, in many ways, redefined how we think about taking care of ourselves. Of course, D-Tech leads to more sedentary work. More of us spend the day typing at our desks rather than moving around. Today, 80% of Americans don't get the recommended levels of exercise (Jaslow 2013).

Yet, we are feeling far more profound *side effects* than these. As discussed in Chapter 3, our social networks and social discourse, in particular, have been re-defined by D-Tech. D-Tech *side effects* are reducing our meaningful social relationships: we're spending more and more time with "the right group for the moment" that D-Tech has helped us find, and creating fewer meaningful connections.

We discussed, for example, how we are each increasingly isolated as a *side effect* of D-Tech. The population of Americans without a single friend in whom they could confide grew more than a third between 1984 and 2004. By 2004, more than half of all Americans (53%) could only confide in a family member; one quarter had no one—*not even* a family member. From 2000 to the present, suicides by middle-aged Americans have increased 50%.

Unfortunately, social relationships can affect mental health, health behavior, physical health, and mortality risk (Umberson and Montez 2010). More concretely, "lack of human connection increases depression, drug abuse, alcoholism, and more" (Lewis 2016)." In fact, the most powerful predictor of how long we'll live appears to be the number of social interactions we have with other people (Pinker 2017). In other words, sitting alone on the sofa with a pint of ice cream—or even a heroin needle—can be a D-Tech *side effect*.

While lack of human connection is increasing, our social discourse with D-Tech isn't helping. Instead of D-Tech doing more to find constructive information about how to maintain our health, D-Tech directs us to conflicting and challenging-to-interpret information. We receive factoids rather than coherent recommendations.

As a result, actionable, attention-grabbing information about food rapidly streams through our virtual world. We can easily find millions of articles discussing the wonders of carbohydrate-free diets, vegetarian diets, and meat-heavy paleo diets. If we combine such recommendations, the only foods we can eat are fruits and vegetables. If we also need to follow a low-glycemic diet to reduce the risk of diabetes, then we can only eat green, leafy vegetables.

We are not rabbits, so something has to give. Unfortunately, each "expert" sounds more convincing than the last and has mountains of facts to support such claims. All of this information is designed to catch our eyes: the information is attention-grabbing, actionable, and easy-to-find. Unfortunately, as a whole, such information fails to help us maintain healthy eating habits. To the contrary, in light of all that conflicting information, more people are simply ignoring all health information and instead choosing to focus on eating tasty, cheap, (unhealthful) foods.

What's the impact? Roughly 1/3 of the American population gained at least 25 pounds between the 1960s and 1980s (Komlos and Brabec 2010), and that weight gain trend is continuing. Almost 40% of Americans and 20% of teenagers are currently obese (Gussone 2017). In summary, D-Tech has done more to help us find destructive rather than constructive advice on maintaining a healthful diet.

While likely a smaller impact today, we should also be aware of the unintended impacts of D-Tech on our environment. As we create a physical world increasingly removed from nature's environment, we may find surprising consequences. For example, consider our ability to disinfect our world. Clearly, reducing germs has a benefit—reduced plague. Nonetheless, almost completely eliminating germs may also indirectly cause serious allergies and other autoimmune diseases. Walk through many American schools, and you'll find signs for "Dangerous peanut allergy" on many of the doors and windows: almost 2.5% of American children today are allergic to peanuts (Gupta 2017). Peanut allergies were rare before 1980 (Hotchkiss 2013).

Without a regular dose of truly dangerous bacteria, our bodies may be focusing our defenses on beneficial foods like peanuts (Velasquez-Manoff 2014). Similarly, indirect and complex relationships may cause the pesticides on our foods to kill bugs in our gut biome—which can also make us sick (Bull and Plummer 2014).

America's lack of health has origins beyond D-Tech. For example, more than 1/3 of Americans don't get enough sleep (CDC 2016). That behavior may be unrelated to D-Tech.

In fact, our healthcare market is successful in many areas. Earlier, we described some of the amazing healthcare technologies being developed and implemented. When we have particular conditions that require treatment, the U.S. healthcare system is, arguably, unmatched. Workers in our healthcare system are some of the brightest and best-educated people in the world.

We're not achieving our purpose for healthcare: efficiently helping us live long and enjoy healthy lives. From only a decade into the Digital Revolution (from 1980 on) we have been throwing more and more resources at healthcare—with little net societal benefit. Today, we spend almost one out of every five dollars in our economy on healthcare. Close up, we see some spectacular innovations and outcomes. However, when we take a step back, we see that the puzzle pieces are not working together. Our answers remain wrong. And, as a result, our health remains poor relative to many countries that spend far less.

D-Tech's complex environment is the underlying cause. We have created a system too complex for humans to fathom, act in, or coordinate. Basic, ugly truths about even common treatments are hiding in plain sight. There's simply too much noise for these ugly truths to receive their deserved attention. Lost is the message of our Agricultural Revolution-era hero Horacio, calling across the plains to "eat well, exercise, and stay healthy."

Is it fair to blame D-Tech for U.S. healthcare inefficiency? Other countries also have access to Digital Technology, but they are not (yet) suffering the American disease.

Yes. The negative impact of D-Tech on our health is *indirect*. In the U.S., D-Tech increased the complexity of an already highly complex healthcare system. It accelerated development of unhealthy habits that were already beginning to develop, such as an increase in obesity (Komlos and Brabec 2010). It hid abuses of the system in a part-private, part-public system involving masses of people. And, recently, the healthcare industries in several other developed countries have been today exhibiting the unhealthy U.S. pattern of the 1980s. Japan, Sweden, and Switzerland, for example, have now experienced multiple years of rapid cost increases with limited improvement in life expectancy. These patterns could certainly be the result of other pressures or, could D-Tech be driving these other countries' healthcare systems closer to the critical levels experienced in the U.S.?

For those following the text of this book, we now turn from our healthcare market to our housing market.

- Financial Market *(page 250)*
- Healthcare Market *(page 255)*

HOUSING MARKET—16% OF U.S. ECONOMY (GDP)

Our housing market, 16% of U.S. economy by GDP (Logan 2017), is failing. Consider the purpose of this market: to deliver reasonable-cost, high-quality housing in the places we want to live—i.e., the places where new jobs exist.

In the 1960s, a cultural shift began that set the stage for the housing market's failure: we began expanding the rights of property owners to block new housing. This movement fully prospered as the D-Tech *side effects* began to impact our environment in the 1970s:

- As discussed in Chapter 4, the volume of regulations exploded, with many creating hurdles to new housing construction.
- Neighborhoods began to suffer from lack of civic engagement, as discussed in Chapter 3. This lack of engagement has formed a barrier to creating a neighborhood coalition that could overcome such regulatory hurdles.

Figure 5.14 summarizes these negative impacts.

As a result, we're not building housing where we want to live. Want to live close to New York? Unless you're wealthy, your future is likely "couch surfing," "faking a real address for potential employers," and "lots of [cheap] Frito's from the bodega" (Bowling 2017). Housing is just too expensive.

Today, we build far less housing than needed in the most successful, best-educated parts of the country (Glaeser 2017).

In other words, D-Tech's hotspots generally have deeply failing housing markets, including New York, Boston, Palo Alto, San Francisco, and Silicon Valley (Glaeser 2014; Hsieh and Moretti 2017; Naimat 2016).

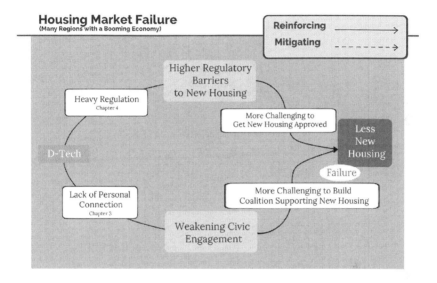

Figure 5.14: Current failures in the U.S. housing markets are side effects of D-Tech: through D-tech, regulatory barriers to new housing are growing, while the civic engagement that could allow us to overcome such barriers is weakening.

According to Hsieh and Moretti (2017), between 1964 and 2009, failures in the U.S. housing markets doubled. This doubling might have cost the United States as much as an additional 50% more economic growth, as illustrated in Figure 5.15.

The left column illustrates the size of the total U.S. economy (everything we produced) in 1964. The right column illustrates actual growth and lost potential in 2009. This value lost could be

40% of 2009's total economy—an amount even larger than the entire 1964 American economy.

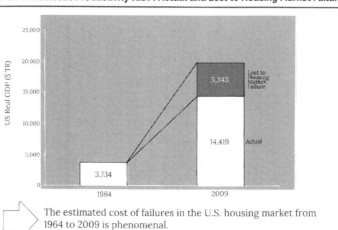

U.S. Gross Domestic Productivity (GDP) Actual and Lost to Housing Market Failure

The estimated cost of failures in the U.S. housing market from 1964 to 2009 is phenomenal.

Source: Calculation based on Hsieh and Moretti 2017, and https://www.thebalance.com/us-gdp-by-year-3305843

Figure 5.15: Housing is a national, rather than just a local issue. The chart above illustrates the economic growth lost because we were unable to bring enough people together in American economic hubs (based on Hsieh and Moretti [2017] and Amadeo [2018]).

Why could the implication of housing market failure be so big? According to Hsieh and Moretti (2017), the failure in a regional housing market has national implications. Figure 5.16 illustrates. The top row illustrates what happens in an innovative, booming region. The economy in Silicon Valley, for example, began to boom because of D-Tech innovation. People wanted to move there for jobs and to be part of the excitement. If the housing market was functioning well, then new housing would be built to meet this new, higher demand. The growing population would fully develop the innovative ideas in the area, and the entire nation would benefit from that innovation.

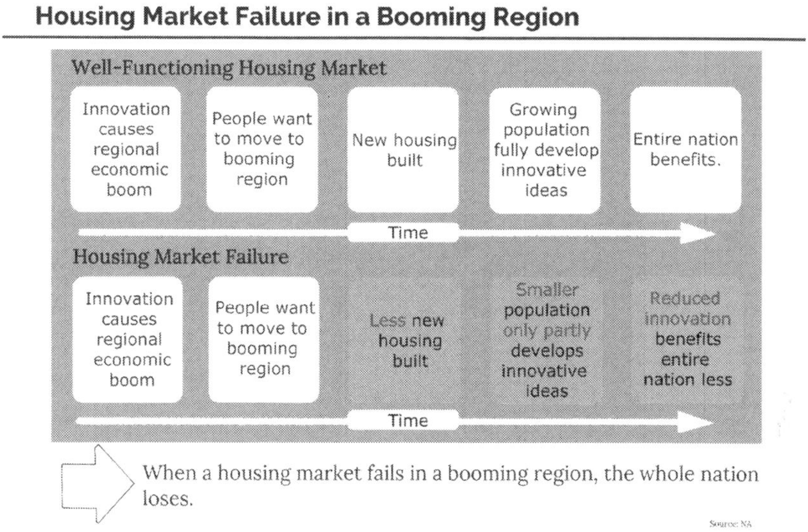

Housing Market Failure in a Booming Region

When a housing market fails in a booming region, the whole nation loses.

Source: NA

Figure 5.16: Housing may be a national issue due to its impact on innovation. With fewer people living in the most innovative regions, less innovation occurs, which impacts the entire nation.

A housing market can fail in a booming region, however. This situation is described on the bottom row in Figure 5.16. People still want to move there, but there simply isn't enough new housing built. The population growth is limited by the amount of places for people to live. And, with a smaller population, the region's innovative ideas are only partly developed—and the entire nation misses the benefits of that lost innovation.[10] Silicon Valley (in California) and Boston, for example, are currently sources of amazing ideas. To get the most out of those ideas, we need people living there who can expand on those ideas and make the most out of them. Unfortunately, the housing markets in both Silicon Valley and Boston are failing to deliver enough housing.

What's the role of D-Tech? As D-Tech *side effects* began to impact our environment, Government hurdles to new housing exploded, while neighborhoods increasingly lacked the civic engagement to get past those hurdles.

Consider the case of California. In 1978, nearly two-thirds of residents voted to cut and then freeze residential property taxes (California Tax Data, n.d.). For those growing up in California, low-property taxes were great in some ways. Nonetheless, this tax structure had unintended consequences: such taxes reduced the desire for municipalities to build new housing. Instead of housing, cities zoned property for more highly taxable commercial use (Dowall 1982). Meanwhile, Government enacted more regulations. Veto rights over new projects were effectively awarded to "a dizzying array of abutters and stakeholders (Glaeser 2017)."

D-Tech did not determine which rules to create. Yet, as discussed in Chapter 4, only with D-Tech could we hope to maintain today's massive web of federal, state, and municipal regulations. D-Tech invited us to create more legislation, and we accepted that invitation with open arms. As American culture began to increasingly support those who owned property, many such regulations were written to provide vetoes on potential new housing construction.

D-Tech also played a key role in making such regulatory blockades more difficult to bypass. In any economically vibrant community, many people actually want new housing. Businesspeople, for example, see the potential to profit by building more houses. Renters (approximately 36% of Americans [Cilluffo, Geiger, and Fry 2017]) love added housing: such housing adds new options—and the added competition lowers rents. Business owners want to draw new employees to their businesses, and sufficient housing helps that effort.

Nonetheless, just as in battle, getting past a blockade requires coordination among like-minded people. Interested renters and businesspeople must have a "Paul Revere" alarm to raise the townspeople:

> *One if by land, and two if by sea; And I on the opposite shore will be, Ready to ride and spread the alarm Through every Middlesex apartment and business, For the neighborhood-folk to rise up and take arm.* (Based on the poem by Henry Wadsworth Longfellow, Paul Revere's Ride)

Even more important than a single activist is the spirit of civic engagement. An army of engaged citizens must be ready to take arm in their self-interest, like the American Revolutionary Army of the 18th century. On the other hand, as discussed in Chapter 3, D-Tech weakened community networks. More people have been emailing friends in Zaire and fewer people have been investing in relationships with neighbors.

The California state government sent a Paul Revere to try to promote new housing. Sacramento officials sent The Metropolitan Transport Commission (MTC) to pressure cities to rezone more land for housing. Yet, no great Revolutionary Army was inspired to support this cause. The army of renters and businesspeople who would naturally want more housing were not sufficiently civically engaged to join the cause. Instead, a few citizens effectively tried to throw bottles at the "socialist" MTC attempting to centrally plan their towns. These efforts largely failed (Grabar 2015).

Through most of history, local coalitions of builders and interested citizens have found creative solutions to housing problems: art projects, support of local initiatives, and the like have often played roles in the solution. New York enacted its pioneering zoning codes in 1916, yet today's housing crunch became a material issue only in the 1970s (Glaeser 2017). Unfortunately, building a local network of concerned citizens interested in an agenda such as housing is challenging to form today.

It may seem odd to discuss volume of housing in terms of D-Tech, regulations, and civic engagement. Doesn't a lack of space limit housing?

No. In most areas, lack of land is not a barrier to new housing. Even metropolitan San Jose, in the heart of Silicon Valley, records only 5,200 people per square mile. That number is miniscule compared to the 26,400 people crowded into each square mile on Manhattan Island (Glaeser 2014). Similarly, the coveted Middlesex County, Massachusetts that includes Harvard and MIT, houses only about 1,800 people per square mile. Even the population density of Harris County including Houston, at about 2,000 people per square mile, is higher.

Where lack of space is an issue, we frequently have the technology to increase land. Many of the areas where we want to live are right on the water. Think of the Netherlands—where 26% of the country is reclaimed land below sea level (Reuters 2010). We could certainly reclaim land in some parts of the U.S. as well. In short, our lack of housing is a failure of the housing market, not a lack of land.

Our housing markets have not failed in all dimensions. Our housing is high quality. Our doors are sturdy, electrical wiring is generally safe, and we are well protected from all but the fiercest storms. The cost of housing is even reasonable in many parts of the United States.

Nonetheless, our housing market can only be described as failing. It is failing in the most critical regions of our nation—those that are rapidly innovating D-Tech. In these areas, Businesses have failed to overcome the complexity associated with D-Tech *side effects*. They have failed to find the nuanced, creative approaches necessary to get critical housing built.

For those following the text of this book, we now turn from our housing market to our food markets. You may also choose to read these sections out of order, or simply to skip to our discussion on <u>Risks to Our Future</u>.

FOOD MARKET—6% OF U.S. ECONOMY (GDP)

Our food markets—comprising 6% of the U.S. economy by GDP (Economic Research Service 2017)—are failing. The purpose of our food industry is to nourish us efficiently. This market is getting the "nourishing" part wrong.

D-Tech's *side effects* are a major contributor to our less nourishing food, as illustrated in Figure 5.17:

- As the Internet provides an increasing muddle of information about health, some of us are simply giving up. We are clear about price and taste, so we buy cheap and tasty food, even if that food is unhealthful.
- Those of us who have not given up on our health are increasingly dependent on labels associated with healthfulness such as organic and low fat. Such labels, however, are imperfect in reflecting healthfulness.
- Meanwhile, we've accelerated our ability to modify our foods. Nourishing aspects of our food not clearly labeled are likely to be lost or weakened.

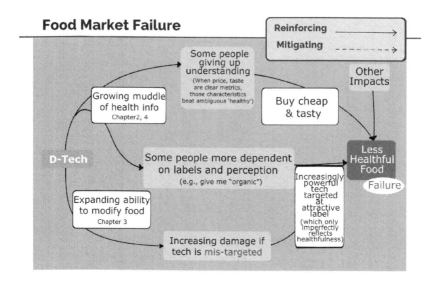

Figure 5.17: Current failures in the U.S. food markets are side effects of D-Tech: through D-Tech, we've confused people and are increasingly mistargeting increasingly more powerful food modification technology.

So, what happens when you ask, "Hey, Hon. Can you please stop at the store on the way home? We need some crackers for the kids' lunches. Get the healthy ones, ok?"

In many cases, we are likely to select sugar-laden crackers baked from ingredients less healthful than their 1950s counterparts.

To review just a handful of the impacts mentioned previously:

- Fruits and vegetables have less protein, calcium, phosphorous, iron, and riboflavin than they did in 1950 (Scheer and Moss, n.d.).

- A chicken in 2004 contained more than twice as much fat as in 1940, a third more calories, and a third less protein (Purvis 2005).

- Apple varieties include more high-sugar types of fruit such as Pink Lady, Fuji, and Jazz (Torrens 2017).
- Organic food may be moderately more nutritious than non-organic today—but it's still not up to the level of the traditionally farmed food that we had in the 1940s, 50s, and 60s (Plumer 2015).

D-Tech accelerated our ability to change our food. We're now breeding a generation of cows every seven or so years. We're breeding, feeding, genetically modifying, concentrating, processing, and chemically reformulating . . . *fast*. Unfortunately, we're accelerating the process of creating less healthful food.

As the noise around healthfulness increases, the results are worsening. More people are giving up trying to identify what is actually good for their bodies.

Not everything is going wrong in our food markets. Our food is, by historical standards, inexpensive. Today, we spend less than a fifth of a household budget on food, versus almost half in 1900 (Thompson 2012). Nonetheless, such cheap prices cannot completely compensate for the failure in our foods' healthfulness.

In summary, our food markets are failing. The purpose of our food markets is to efficiently nourish us—and we are not being nourished. We have amazing technology with the potential to produce more healthful food. We might even bring the healthfulness of our food and our diets back to the level of the 1960s. Shoot, we could probably do even better than that. Nonetheless, for the most part, such "healthful food" technology remains largely untapped.

For those following the text of this book, we now turn from our food markets to our education markets. You may also choose to read these sections out of order, or simply to skip to our discussion on Risks to Our Future.

EDUCATION MARKET—6% OF U.S. ECONOMY (GDP)

Our education market, also 6% of the U.S. Economy by GDP (National Center for Educational Statistics 2016a), is failing.[11] The purpose of education includes efficiently accelerating our learning:[12] learn with a few years in a classroom what would take decades to learn without the classroom's support. Through this learning, education should help prepare our citizens, especially our youth, to participate successfully in society.

Since 1970, education has made little progress in educating our youth despite our allocating more than twice as many resources to that purpose.

As illustrated in Figure 5.18 below, D-Tech's *side effects* are a major contributor to the growing inefficiency and failure.

- Through D-Tech, we have created an increasingly complex educational system. Continuous efforts to improve that system are not delivering better outcomes—they are getting lost in the complexity.
- D-Tech *side effects* are causing many neighborhoods to deteriorate (as discussed in Chapter 3), which increases students' challenges in learning.

We will explore these links over the coming pages.

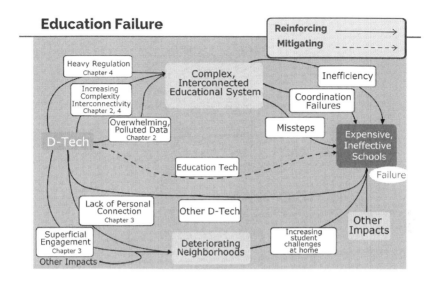

Figure 5.18: Current failures in the U.S. educational system are side effects of D-Tech: through D-tech, we have increased the complexity of our educational system, while too many neighborhoods (the primary areas where our youth learn) are deteriorating.

Let's begin with the current status of U.S. education. Perhaps Vivian Cox Fraser summarized it best: "Everyone's getting paid, but Raheem still can't read (Russakoff 2015)." We're spending more than twice as much per pupil on education as we were in 1973. Notwithstanding, those additional resources aren't delivering better outcomes. Figure 5.19 summarizes the situation.

Let's look at primary and secondary education first. We've increased primary and secondary school spending per-student by 111% since 1973: an increase of more than $5,000 per student per year. Unfortunately, such spending has had little impact on our children's learning. The scores of our 17-year-olds on a standardized math and English test (the NAEP) are highlighted in Figure 5.19. These scores are essentially unchanged. In 1973, the average American 17-year-old scored 286 points. In 2012, this was up to 287 points. In mathematics, the average American 17-year-old scored 304 points in 1973. In 2012, they scored 306.

U.S. Education Cost and Standardized Scores

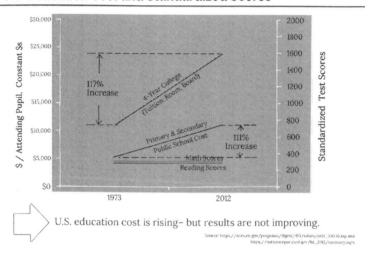

U.S. education cost is rising– but results are not improving.

Source: https://nces.ed.gov/programs/digest/d13/tables/dt13_330.10.asp and https://nationsreportcard.gov/ltt_2012/summary.aspx

Figure 5.19: We've more than doubled spending on both lower and higher education per pupil in the U.S. during the D-Tech era. Improved outcomes, however, are not forthcoming (based on National Center for Educational Statistics [2016] and [n.d.]).

If we add all the additional spending over one student's career through high school, we're spending about $60,000 more per student today than we were in 1973. That's a lot of money. For that additional investment, our NAEP exam scores have risen 0.5%. Let's try a word problem associated with that:

"Teacher Hegenbarth, suppose we wanted to change our letter grade by one level. In particular, suppose we would award our schools a grade of 'D' today but wanted to raise that to a 'D+'. How much would that '+' cost us at our current rate of improvement?"[13] Teacher Hegenbarth responds, "At the current rate, that '+' would cost $360,000 per student." Teachers often require a 3% improvement before they reward such a higher score, and each 1% improvement is currently costing us $120,000 per student today."

Ms. Hegenbarth pauses for a second. "That '+' would pay my salary for *many* years!"

Measuring education's results is notoriously challenging. However, the NAEP exam scores seem to tell the basic story. Our limited ability to transfer what we learn from school to the real world is consistent with our mediocre NAEP exam scores. Students remain largely unable to use the following first lesson to solve the second challenge:

- **Lesson Problem:** An army needs 100 soldiers to successfully storm a fortress. The land around the fortress is mined so that any force with more than 25 soldiers will explode.

- **Solution (provided to the student and illustrated below):** Divide the force into five groups of 20 soldiers. Each group attacks from a different side, meeting at the fortress.

- **Challenge problem:** You are a doctor responsible for a patient with a malignant tumor. The only way to kill the tumor is 100 units of a special killer ray. Unfortunately, any part of the patient's body that receives more than 25 units will also die. How do you kill the tumor without killing the patient?

Sample Desired Answer: Target the tumor from five different directions. Fire 20 units of the ray at the tumor from each direction.

Only about one in five students is able to solve the challenge problem—even after being provided the lesson problem's solution.[14] We don't have equivalent data from 1973, so we're unsure if that's an improvement. Nonetheless, lack of comparison data isn't critical—even if that result is 100% improvement, such improvement would be pathetic. *We're spending $60,000 more on every student through high school*—and these students can't recognize a simple, logical pattern! Moreover, as we will discuss shortly, a growing number of our youth are having trouble integrating into adult society. It's hard to argue that our public grade-school education has materially improved.

Figure 5.20: Both top and bottom problems are conceptually identical: to storm a fortress when you cannot attack with all your troops from the same direction, split your forces. Yet, most students were unable to solve the bottom challenge even after being provided the answer to the top one.

Now let's look at four-year colleges, taking another look at Figure 5.16.

The cost increase for colleges is similar to the cost increase we saw for secondary education. After adjusting for inflation, the cost of a four-year college education increased 117% from 1973 to 2012 (National Center for Educational Statistics 2016b). The average four-year college attendance today costs close to $25,000 per year. For Michigan residents, for example, University of Michigan (including tuition, room, and board) costs $29,526 per year (1st Financial Bank, n.d.). Small, liberal arts college Wesleyan University costs $66,640 per year ("Wesleyan University," n.d.).

Despite such expenses, too many students lack fundamental skills. "At more than half of the [public universities tested], at least a third of seniors were unable to make a cohesive argument, assess the quality of evidence in a document, or interpret data in a table (Belkin 2017)." A large number of students showed no improvement in these skills over their four years of college and, at some universities, students even seemed to regress (Scarola 2017). Once again, we don't have a clear comparison of what students learned in 1973 and what they learn today. Nonetheless, we still know enough to call the *additional* $50,000 spent over the average college student's four-year career a failure.

When considering the size of this failure, it's useful to consider student debt. Many American youth borrow money to pay for such expensive (yet too frequently underwhelming) education. To support college students, the U.S. Government (and its taxpayers) have guaranteed that they will pay the debt if the students are unable. We, Government taxpayers, have now guaranteed $1.4 trillion in outstanding student debt. That is a guarantee of more than $4,000 for every man, woman, and child in our country. (In other words, my family of five is, on average, responsible for the guarantee on $20,000 in existing student loans. Wow!)

Unfortunately, many students graduate from college without the earning power needed to repay their loans. A full 61% of the student debt in the U.S. is not currently being repaid: 17% is delinquent and 44% is currently postponed ("A Look at the Shocking" 2018).[15] If students do not pay off their loans, Government will have to.

D-Tech *side effects* were a major influence in these failures. As we've discussed throughout the past three chapters, as we apply D-Tech, we are creating increasingly complex systems: rapidly changing, many-optioned, interconnected systems about which there are masses of polluted information. As discussed in Chapter 3 and earlier in this chapter, D-Tech *side effects* have also played key roles in our neighborhoods deteriorating.

American schools are complex. D-Tech provided us the tools with which to create more complex regulations and school systems. We've taken full advantage. Everyone seems to have an idea about how to improve our schools, and that has led to intricate, interwoven layers upon layers of regulations. Local school boards set rules concerning health and safety. States determine the curriculum taught. The federal government adds regulations targeted at improving results, maintaining oversight, and ensuring everyone is fairly served.

D-Tech did not write the mountain of rules at the local level, state level, and federal level. Yet, as discussed in Chapter 4, only with D-Tech could we possibly hope to maintain today's web of regulations. D-Tech sang a siren song, and we ran to it—creating a mass of legislation. Unfortunately, while D-Tech helped us create that legislation, it *doesn't* help us understand the resulting complex system.

To understand the impact of such regulation, consider the exhausted 1st grade teacher, Ms. Hoag. Ms. Hoag teaches a class of 25 children. Two students require an intervention plan because of bad behavior (Ginny and Thom), and five are behind grade level standards. Here I describe only some of the paperwork the real Ms. Hoag explained to me.

By 5:30 p.m., Ms. Hoag has completed the lesson planning. She knows what she'll teach tomorrow, and has referenced that lesson with her state's 86-page curriculum guide. The official requirement is awkward, but she's confident she has it nailed: "2.2 Comprehension and Analysis of Grade-Level-Appropriate Text—Use pictures and context to make predictions about story context" (California Department of Education 2000).

Now Ms. Hoag begins on the regulatoryt challenges:

- Ginny and Thom: For Ms. Hoag, Ginny and Thom are almost as challenging as the paperwork she needs to complete because of them. For each behaviorally problematic student, Ms. Hoag must enter notes about the child's behavior and Functional Behavioral Plan every day. She's thankful she's done with the two-week, time-intensive process of daily note-taking and plan-writing by which she created the plan. That entire process was mandated to her.

- For her five behind-grade-level students, Ms. Hoag also records her daily notes. It takes her an hour. Once again, Ms. Hoag's thankful she finished earlier in the year the preparation of these student's Intervention Plans—composed of 12–18 different forms each.

- Tomorrow morning she'll prepare the bi-weekly computerized testing she needs to give each of these children. Ms. Hoag will need to help each of these students—who are behind first grade standards—enter the 9-digit school ID number, 5-digit student code, and 4-symbol test ID number.

- School newspaper: Ms. Hoag enjoys coaching the older children for a change of pace. However, freedom of speech is one of the most litigated areas of school education today. The kids want to write an article about the impact of divorce on learning in school. The topic is a wonderful challenge for the kids—but it could too easily become a legal challenge for Ms. Hoag. The First Amendment (free speech) exists in a more limited way for schools and there is a mountain of legal precedent ("Education Law" 2018) about which Ms. Hoag needs to be careful. She writes an email to the school's attorney asking for more background.

Ms. Hoag is not finished with all this until 9:30 p.m. The last three hours of her day were focused only on paperwork and legal issues. Certainly, some legal issues existed before D-Tech. However, they were more limited. In 1964, for example, only eight pages of federal regulations were dedicated to education. By 1976, that number was 360 (McGuinn 2015). The bulk of the regulations and precedence worrying Ms. Hoag tonight were created with the assistance of D-Tech.

Today, there are so many regulations that "it is probably fair to say that there is not one [educational] institution that is able to be in complete compliance with all of the federal laws" (McMahan and Loyola 2011, 14–21). The direct cost of complying with the regulations has certainly played a role in increasing cost. For example, a "study from Vanderbilt University found that nearly $150 million, or 11%, of the school's total expenditures in 2014, were devoted solely to complying with federal regulations" (Carp 2015).

There's even more. A school is an integral part of a neighborhood. As we discussed in Chapter 3, D-Tech *side effects* have played important roles in multiple aspects of the deterioration of our neighborhoods. Consider these impacts:

- The weakening relationships with our neighbors and those physically close to us.
- The increasing challenge of aligning our Culture's contradictions, including between our treatment of women and children. As a society, we began to prioritize women's rights over the nuclear family decades ago. However, we're still miles from developing a Culture consistent with such priority. If women's rights are to be a priority, single parents need consistent support to raise their children well. Certainly many single parents succeed in raising their children well, but they must overcome enormous challenges.

Shortly, we'll discuss further D-Tech *side effect* that have also played important roles in the deteriorating of our neighborhoods. These are the *side-effects* of leaving millions and youth unemployed.

As a result of such neighborhood disintegration, school success is more challenging to achieve. Russakoff (2015, 214) describes a struggling teen pushed over the edge by violence:

> *In March, a lifelong friend of Alif's was stabbed and killed at age fourteen while the two of them walked home from a pickup basketball game. It was a calamity that could have derailed any child. For one already at risk, [this death] was all the more devastating. For weeks, his mother said, Alif cried himself to sleep and lost all motivation. He ended the year with F's in English, math, and history, and was sent to summer school.*

In the city of Newark, New Jersey, 71% of children come from single-parent homes and 41% live below the poverty line (Russakoff 2015, 88). In some parts of the city, as many as one-third of the students are forced to move from one school district to another every year (Russakoff 2015, 202). Childcare is challenging to find and difficult to pay for.

The job of a school is to help a neighborhood raise its children. Only if the school is able to manage the challenges of that neighborhood does it have a chance at success.

Certainly there were tough neighborhoods before D-Tech. However, as D-Tech combined with other stresses to challenge our neighborhoods, school success became harder to achieve.

D-Tech has brought us the masses of data we love. However, such data can provide an obscuring mask, as well as clarifying insight. During the 1960s, data and regulation on schools were focused on where money was spent and how bureaucratic processes were monitored. More recently, data has been focused on curriculum and teaching methods (McGuinn 2015). Both types of data are challenging to interpret—and both increase the regulatory burden.

Moreover, records do not necessarily equal reality. As Russakoff describes:

> One of the cruel ironies of Newark's schools was that throughout their long decline, the district often appeared on paper to be in perfect compliance with all requirements. A 2009 outside review of the Avon Avenue School found that the curriculum aligned well with state standards in math and literacy. And the district had a written policy requiring principals to observe teachers regularly, supplying feedback and coaching to ensure they were reaching all children (Russakoff 2015, 48).

Reality, on the other hand, bore little resemblance to that script:

> An application for federal aid, filed shortly after the outside review, reported that too many teachers displayed "an inability to captivate student interest and motivate them." And there was "minimal use of higher order questioning." The quality of the curriculum had little relevance without effective instruction (Russakoff 2015, 49).

It is even more challenging to obtain meaning from data because schools are integral parts of a neighborhood. Only by including information about the neighborhood can a meaningful analysis be done. However, we must understand the neighborhood before we can collect meaningful data.

With such complexity, it makes sense that effort after effort to reform our schools fail. Consider, for example, the $200 million effort to reform Newark, New Jersey's school systems. Mark Zuckerberg, CEO of Facebook, donated $100 million on condition that a matching $100 million could be obtained. Mark and his team wanted to totally transform Newark schools: introduce charter schools to replace the worst-performing schools, provide higher pay for teachers who performed better, and modernize the use of D-Tech to analyze students' progress.

The attempt to transform Newark schools failed. Scores fell across the district, as well as in all but one of the schools designated as needing urgent renewal (Russakoff 2015, 203).

Some of the mistakes made during the reform efforts were independent of D-Tech. In particular, the superintendent in charge did not involve local residents when planning the massive upheaval she created. Even the announcement of her vision came at a cocktail party with few locals in attendance (Russakoff 2015, 198). Locals whispered about, "How Ms. Anderson [the superintendent] could announce such massive changes without neighborhood leaders."

Nonetheless, ignoring the neighborhood issues serves only to highlight the complexity of the challenge. The superintendent focused only on the school system itself. Despite that over-simplification, she talked about, "having to play 16 dimensional chess" in her planning. She simultaneously tried to answer many questions, including:

> *How to improve district schools fast enough to persuade fami-*
> *lies to stick with them. How to close underpopulated schools*
> *without adding to neighborhood blight. How to retain the best*
> *teachers, given that she estimated she would have to lay off a*
> *thousand of them in the next three years. How to find money*
> *to modernize schools that on average were eighty years old.*
> *How to stabilize the district's finances for long-term survival*
> *(Russakoff 2015, 200).*

Once the plan was announced, the additional challenges of collaborating with the neighbors became clear. A parent asked Ms. Anderson, "'Can you guarantee my daughter's safety?' The parent went on to describe gang activity and drug dealing near the school where her daughter would be reassigned (Russakoff 2015, 202)." The problems went deeper:

> *Neighborhood schools were part of a delicately balanced*
> *ecosystem that served many needs for families. Anderson*
> *persuaded charter schools to take over three neighborhood*
> *K–8 schools, portraying the move as an opportunity for some*
> *of the lowest-income children to attend the most sought after*
> *programs. But the charters agreed to serve children only in*
> *kindergarten through fourth grade. Children in the upper*
> *grades had to go elsewhere. That removed a trusted source of*
> *childcare for parents who relied on older siblings to accompany*
> *younger ones to and from school (Russakoff 2015, 202).*

These attempts to manage such complex change inevitably led to missteps as well.

With the district's attention focused on implementing sweeping change, some of the most basic functions—such as student scheduling at certain high schools—had broken down. Throughout the fall semester, Alif had no English or math classes. Although Anderson's Grad Tracker program was supposed to ensure that students who failed classes regained the credits, there was no arrangement for Alif to make up freshman English and history (Russakoff 2015, 215).

Inefficiency in such complex systems, both during major transformations and beyond, is rife. Money gushes and oozes in a myriad of directions. A colleague described a school janitor who must spend eight hours cleaning a local gym that only requires two hours to clean. In 2009, a district audit in Fort Worth found that more than $1.54 million was overpaid to district employees and staffers; $2.7 million worth of computer equipment and technology was either "unnecessary" or was unused in its first seven years after purchase (Berard 2014).

Certainly, there are efficiently run school systems. Some of the best-educated students anywhere come out of American schools. However, the never-ending series of privatizations, regulatory changes / additions / or modifications, funding revisions, and teacher improvement exercises is not working.

The result of the attempted transformation in Newark, New Jersey, likely describes the result of many such efforts. Data were gathered. Consultants provided reports. Money gushed and oozed. "Everyone's getting paid, but Raheem still can't read."

Not every problem in American schools is necessarily a D-Tech *side effect*. Perhaps the recent, horrific violence has origins other than D-Tech and its *side effects*. Umair Haque (2018) suggests we've had 11 school shootings in the U.S. during a 23-day period. Such shootings could be the result of the D-Tech side effects on our Culture described in Chapter 3—or they could be from a completely separate origin.

It's also possible that some of the growing university spending may not be associated with one of the D-Tech *side effect*. A material portion of spending growth is associated with how college students are spending their time. A recent study of

University of California undergraduates (Brint and Cantwell 2010), for example, reported that students were spending:

- 13 hours a week studying
- 43 hours a week on recreation[16] Many universities are increasingly focused on such recreation. University dining experiences are described:

> "With more than 500 staff members on UCLA's dining team, it can't be surprising that the university is committed to making sure their students receive the best collegiate food experience possible (Kaufman 2013, slide 56)."

> "Bowdoin College... serves dishes like mussels in butter sauce, haddock with jalapenos, and roasted root vegetables with polenta." Bowdoin is also well known for the yearly lobster bake it holds for graduating seniors (Kaufman 2013, slide 60).

Of course, the focus on luxury is broader than just on food. "A free movie theater. A 25–person hot tub and spa with a lazy river and whirlpool. A leisure pool with biometric hand scanners for secure entry. A 50 foot–climbing wall to make exercise interesting. . . [These are] facilities you can find on a college campus (Newlon 2014)."

Spending on teaching, itself, is increasingly on the sizzle rather than the steak, as well. Digital blackboards are a booming business (Mo 2017). Despite such spending, universities are investing less in the teachers using those blackboards. In 1969, 78% of teaching positions at colleges were tenure-track professors. In 2009, that number had shrunk to 33% (Kezar and Maxey 2013).

It's not that universities are shifting teaching roles to technology that replaces teachers. The so-called Massive Open Online Courses (MOOCs) are currently playing only small roles in teaching at universities (Pope 2014). Increasingly, universities are having teaching provided by part-time instructors rather than the best and brightest in the field. Certainly some of these non-tenure track teachers are excellent, while some tenured professors are horrible teachers. Nonetheless, this teaching model does not feature education by the stars in the field. Meanwhile, as we discussed in Chapter 3, the average academic research isn't setting records for quality, either (Eisen, MacCallum, and Neylon 2013).

Most students don't want to borrow thousands of dollars for a few years of rock climbing and grilled haddock. Perhaps there are exceptions; most likely *some numbers* of college students are attending college for the country club experience.

On the other hand, more students attend college because doing so is increasingly required for today's job market. What percentage of college students leave university with debt? 70%. For 40 million Americans, student debt still hangs over them (Berman 2016). For such people, a school with a bit less haddock and a bit better teaching would undoubtedly be attractive.

However, whatever we say about university lobster bakes and leisure pools, three issues still remain:

- Especially as D-Tech progresses, American society today needs graduates who are better, brighter, and more effective at supporting society.
- Our American educational system is increasingly failing. We spend more than twice as much per student on education today as in 1970, yet our high school graduates remain unable to apply basic logic. (Even large portions of our college graduates cannot follow basic logic.) A major portion of our college graduates can't pay off the student loans used to obtain their education.
- D-Tech *side effects*—an increasingly complex school system and deteriorating neighborhoods—have played key roles in creating these failures.

Is it fair to blame D-Tech for U.S. educational inefficiency? Other countries have access to D-Tech, but many countries have more consistently successful schools.

Yes. The negative impact of D-Tech on our education is *indirect*. In the U.S., D-Tech increased the complexity of an already highly complex educational system. Compared to the U.S., most nations have fewer students, layers of regulation, and degrees of uniqueness in their educational systems. The Culture of every nation has also interacted uniquely with D-Tech.

It's quite possible some countries have found a way to avoid the complexity problems rife in American education. Nonetheless, D-Tech has almost certainly led to increasing complexity in most countries' educational system. The stories of such complexity and their impacts would definitely make interesting complements to this discussion.

For those following the text of this book, we now turn from education to the grey market: business interactions that do not officially exist. You may also choose to read these sections out of order, or simply skip to our discussion on <u>Risks to Our Future</u>.

GREY ECONOMY—8% OF THE U.S. ECONOMY (GDP)

The grey economy, 8% of the U.S. economy by GDP (Schneider 2012), is a failure by definition. The grey economy refers to work done without required licenses. Grey economy businesses take in money and deliver products and services like any other business. However, these are unlicensed. As a result, Government does not play its role in supporting grey economy interactions. Instead, Government tries to shut such businesses down.

As licensing and other Government requirements expand in our D-Tech economy, more Business efforts are relegated to the grey economy.

Measuring the size of the grey economy—which officially doesn't exist—is challenging. Yet, it appears the share of our economy representing grey-market Business has approximately doubled since the start of the D-Tech revolution. During the 1970s, the grey economy grew to approximately 4.5%–6.1% of the total economy (Murray 2017). This segment expanded again between 2007 and 2011 to approximately 8%–10% of the total economy (Schneider 2012).

Grey economy businesses are similar to the business in the story below (based on Barnes [2009]):

> Laurie runs along Harlem streets, her cart filled with tube socks. She asks passers-by to purchase a pair of socks and then runs on. Laurie has to keep moving to avoid the police, as she doesn't have a valid vendor's license.

The informal, grey economy is big. It's estimated to be in the ballpark of $1 trillion. That $1 trillion estimate excludes drugs, counterfeiting, and all other activities that are inherently illegal. It includes only sellers like Laurie whose activities would be legal if the sellers had obtained appropriate licenses.

What is the relationship between the grey economy and D-Tech? As the challenges in complying with society's rules increase, more people choose to conduct business outside these rules. Such people give up on obtaining the benefits of Government support for Business such as contract enforcement. When the costs of complexity become too high, those costs outweigh the benefits of official work.

There is nothing inherently complex about registering a sock-selling business. Yet, as we continue to apply D-Tech, increasing complexity becomes a *side effect*. Government needs paperwork for everything. Even the paperwork for a sock business becomes complex, and Laurie gives up.

Some aspects of the grey economy are positive. A handful of creative, freelance professionals make a solid living from the grey economy (Rouse 2016). People transitioning to the workforce such as teenage babysitters may also obtain constructive experience. Even so, the bulk of people involved in the grey economy are like Laurie—involved because they have no other choice (Barnes 2009).

For those following the text of this book, we now turn from the grey market to our labor market. You may also choose to read these sections out of order, or simply skip to our discussion on <u>Risks to Our Future</u>.

LABOR MARKET—ACROSS THE ENTIRE ECONOMY

Our labor market is also failing. The purpose of the labor market is to match people willing to work with those employers who need work done. Such matching is required across all Business and Government activities. Too frequently, matching does not occur. Or, people end up with jobs that poorly match their capabilities.

As illustrated by Figure 5.21, D-Tech *side effects* are a primary contributor to these challenges.

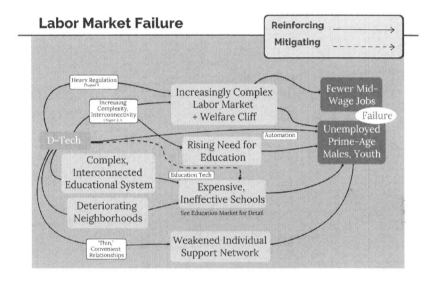

Figure 5.21: Current failures in the U.S. labor markets are side effects of D-Tech: through D-tech, increasingly we're creating fewer mid-wage jobs, weakening education, and decreasing several aspects of our job search networks.

- D-Tech side effects in both Government and Business have increased the labor market's complexity.
- As discussed previously in this chapter, our educational outcomes are not keeping pace with the requirements of our complex D-Tech world.

- D-Tech has played key roles in automating many jobs, particularly for the less-educated.
- D-Tech has played a key role weakening our social ties, making it easier for individuals to drop out of mainstream society (see Chapter 3).

As illustrated in Figure 5.22, these challenges are leading to labor market failures. For many Americans, any job would be better than none. Many of our youth and prime-age males are unemployed. Many workers with mid-level capabilities are ending up in low-wage jobs.

What does the mix of existing jobs look like in the U.S.?

Our jobs, particularly at the lower end of the social spectrum, are warped. Government-driven social programs certainly have a value. Unfortunately, our businesses have not maintained a socially constructive dialog with Government. Take, for example, the "Welfare Cliff" (Baetjer 2016), as illustrated in Figure 5.22.

The horizontal axis is the amount of money Susie earns through traditional employment. This horizontal axis refers to what an employer describes as wages: "Susie, you'll be earning $8 per hour. If you work 40 hours a week at that wage, you'll earn about $16,000 per year." This description corresponds with point A* on the horizontal axis of the figure. The vertical axis shows how much money Susie will actually take home as a single parent with two children—including taxes and subsidies from the government for her rent, electric bill, and more.[17] With a job paying $8/hour, Susie will take home a bit over $60,000 a year (point B* directly above A*).

The basic relationship between earned income (horizontal access) and government support is straightforward. The more money Susie makes in a traditional job, the less the Government helps out. Yet, the full implications of this system depend on the interaction of seven different types of subsidies administered by multiple government agencies. Each subsidy reacts differently to everything from where Susie works, to where she lives, to how many children she has. This complex system could have developed only with the assistance of digital technology.

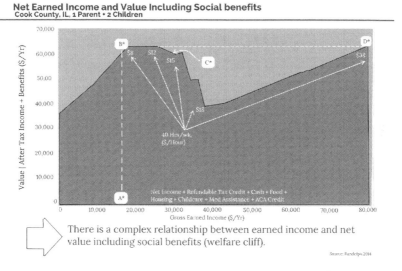

Net Earned Income and Value Including Social benefits
Cook County, IL, 1 Parent + 2 Children

There is a complex relationship between earned income and net value including social benefits (welfare cliff).

Source: Randolph 2014

Figure 5.22: Effective net income for many Americans decreases when salaries rise above modest levels.

In reaction to this complexity, we've messed up the relationship between take-home pay and government support. Suppose Susie is making $9/hour and gets promoted to a $14/hour supervisor's role. Congratulations, Susie!

Unfortunately, Susie's take-home pay just went down with that raise! Her net take-home pay after the raise is point C* on the graph. Every dollar of her raise leads to a more than one-for-one reduction in government subsidies. Susie may get a bigger paycheck. Yet, when we include the impact of Susie's changing paycheck on her government subsidies, the net result could flip—Susie could end up with *less* in total. For example, the Government may stop paying her electric bill and subsidizing her rent and her healthcare. Susie could lose even more in government subsidies than she gains in wages. This relationship is called the "Welfare Cliff." Only if Susie somehow obtains a job paying $38 per hour (point D*) is she as well off as she was at her original, $9/hour job.

The Welfare Cliff creates further *side effects*. The Welfare Cliff makes it more difficult to operate a business with mid-level capabilities. Want to hire a $22/ hour creative salesperson who can come up with creative offerings at a boutique? Forget it. That salesperson may be better off making $10/hour at Walmart. The Welfare Cliff impacts the type of businesses we create and those that survive; it supports Walmart and McDonalds, two companies that pay lower wages. This same cliff makes it more challenging to operate a business that needs more capable employees.

Let's shift from how much people are paid to whether or not people have jobs at all. The U.S. unemployment rate may have been an incredibly low 4.1% in 2015 ("Unemployment Rate" 2018). Economists may say things such as, "Every time I have a conversation with manufacturers or folks in construction, their biggest complaint is they can't get enough workers (Mauldin 2018)." Yet, millions of prime-age, capable people still aren't working. In fact, essentially none of the U.S. jobs created since 2009 are traditional, full-time jobs. Why is that? Who's not working? (See Eberstadt 2016; Saraiva and Matthews 2017.)

Jim and Antoinette aren't working.

Jim stretches his arms and yawns.[18] Jim's just waking up, and he is every bit the bear; it's 1 p.m. in the afternoon. Jim didn't get to sleep last night until 3 a.m., but that's still more sleep than most employed people get. Like 16% of prime-age American males (25–54) without a high school degree, Jim has no job.

TV and the Internet are this bear's tranquilizers. Jim roars at the TV. He reaches for the remote and flicks through the TV channels. He'll watch action movies and M*A*S*H reruns for a few hours and then browse through Facebook for several more hours. Jim's life isn't exciting, but he passes the time. He spends more time on TV and the Internet than anything else in his day.

Jim can work, and does so when he can. However, Jim's knees and ankles hurt constantly; he can climb stairs only with a major effort. Jim hasn't had a regular job for months now, but he sometimes gets gigs laying cement or pounding nails for a few days at a time. He loves to teach the young construction cubs the carpentry tricks he learned when he still thought he'd have a career.

Jim's on-again, off-again girlfriend (Momma Bear) keeps him on the straight and narrow, drug-free, and a decent guy. Knowing that Jim's lost two friends to opioids in the past six months, Momma Bear would despise him if he took drugs. Jim thought one of the friends, Chris, was a close friend but was unaware Chris was using drugs. Jim awoke one afternoon to Chris's mom crying through the phone; she'd found Chris overdosed in the bathtub, wet and cold. It was an opioid death too close to home.

"Jim" is all too common. He represents more than 10 million prime-age, unemployed American males (Eberstadt 2016). These males (age 25-54) tend to be unmarried with no children. TV and the Internet are keeping them busy 5.5 hours per day (Katz 2015). They sleep long hours. They socialize with each other—but for less time than do unemployed women (Katz 2015). They gamble and use tobacco regularly. Half admit to a recent use of an opioid substance (Eberstadt 2016).

The U.S. is 31st out of the 34 developed (OECD) countries when ranked by participation of prime-age males in the labor force (Council of Economic Advisers 2016).

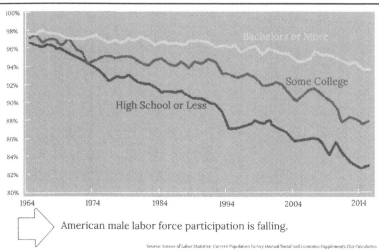

Prime-Age Male Labor Force Participation By Educational Attainment

American male labor force participation is falling.

Source: Bureau of Labor Statistics, Current Population Survey (Annual Social and Economic Supplement), CEA Calculation

Figure 5.23: American male labor force participation
has fallen throughout the D-Tech area. The decrease
is most pronounced for individuals with least
education, but holds at all education levels.

As shown in Figure 5.23,[19] this failure of the prime-age male labor market spans all education levels. In 1964, almost 98% of those prime-age males with a college degree or more were employed; today, that's down to 94%. Labor market participation fell even more dramatically for males with less education. In 1964, more than 96% of prime-age males with only a high school degree were employed; today, that number is less than 84%.

Let's look at another age group. Antoinette is also unemployed. She graduated from high school three years ago in the bottom third of her class. At least she graduated. Antoinette's had a few jobs bagging groceries since then but nothing in the past nine months. She's living in her parents' apartment, playing the online video game League of Legends most days as Stargazer35.

Suddenly, Antoinette knows what she's going to do. The Delta Airlines flight attendant program is singing her name! Her starting salary would only be roughly $25,000 per year and Delta admits, "Work-life balance can be tricky." The job includes health insurance, a 401K-retirement account, and some travel perks. Yet, more than anything, it's the chance at a real job.

Antoinette is one of more than 125,000 applicants for 1,000 Delta Airlines flight attendant jobs. Yes, people applying to Harvard University have a six-times-higher chance of being accepted than do people applying for these flight attendant jobs (Durden 2017b). Even so, Antoinette feels the energy. She sees herself in the polyester uniform already. She breathes in and clicks the "commit" button.

The unemployed youth are Antoinette. One-third of Americans age 18 to 34 now live with their parents. Twenty-five percent of the men living at home are not in school, have no job, and are not looking for one either (Fry 2016; Mosendz 2017). These young folks are called NEETs: not in education, employment, or training ("NEET" 2017).

Are these NEETs just being happy and carefree? Perhaps they're lazy because they can be? No. These youth are not simply enjoying leisure. They are frozen and depressed. NEETs tend to play a lot of video games: many play more than two hours a day. Those who do play more video games report that they feel happier than those who do not (Swanson 2016; Thompson 2016).

What has been the impact of D-Tech in the failures of the labor market? As discussed in Chapter 4, with D-Tech, the spectacular complexity of Susie's government subsidies has exploded. As a result, some people with lower paying jobs would actually be *worse off* if they allowed themselves to get a raise. That's absurd. As a result, we have too many Walmarts and McDonalds. And finding the appropriate, mid-skilled workers such as boutique salespeople and carpenters is a nightmare.

Only through D-Tech *side effects* could we have created such a spectacularly complex, regulatory mess. Susie receives seven different subsidies overseen by multiple government agencies. No D-Tech computer created any subsidy regulations itself. Yet, because we have computers, we have created and stored more and more of them.

As discussed earlier in this chapter, D-Tech has also played a critical role in tossing Antoinette off any realistic life path. Antoinette's education is mediocre and she's missing the social contacts that would help her plan her life effectively (see Chapter 3). Once again, no robot is lying to our children. At least thus far, only a few robots are saying, "Skip your homework and play with me."

Nonetheless, only through D-Tech could we have created today's complex, educational system in which, "Everyone's getting paid, but Raheem still can't read." Most jobs today would require more ability to handle complexity than Antoinette can handle. As a result, Antoinette is stuck trying to compete with a massive army of applicants. They're all competing for $25,000 per year flight attendant jobs with limited potential for a balanced work-life. People chasing Harvard University's brass ring have a better chance of success.

As discussed in Chapter 3, D-Tech has also weakened the connections in our neighborhoods. D-Tech *side effects* of weakened social connections, an increasingly complex world, and a challenging employment market disconnect us with society. This disconnection may also play a role in increasing the likelihood that Jim will have an arrest record. People without social ties are more likely to commit anti-social acts. Over the past 30 years, the number of American men with arrest records has increased 400% (to 14 million)—and people with arrest records find getting jobs more challenging.

Finally, D-Tech now plays a more direct role in the current failures of our job markets. D-Tech has helped limit many of the jobs that once employed less-skilled men. As illustrated in Figure 5.24, between 2000 and 2015, for example, U.S. employment in manufacturing, construction, and utilities fell 24% with much of that decrease due to increasing automation.

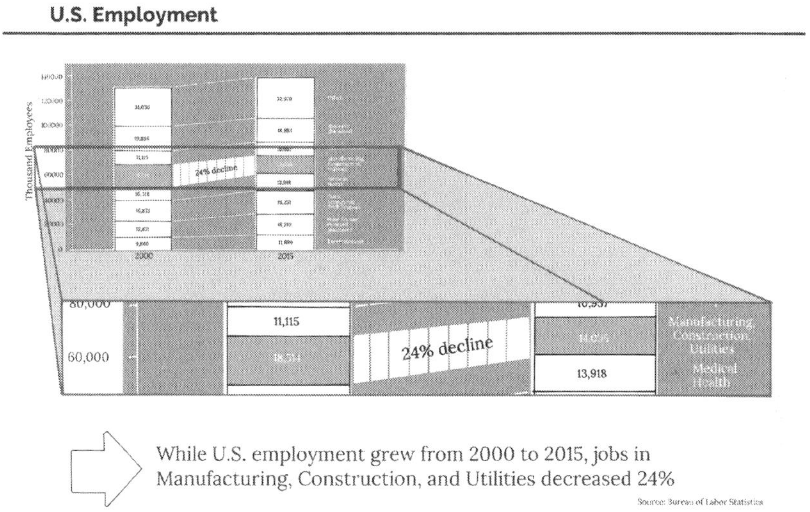

U.S. Employment

While U.S. employment grew from 2000 to 2015, jobs in Manufacturing, Construction, and Utilities decreased 24%

Source: Bureau of Labor Statistics

Figure 5.24: American manufacturing jobs have fallen by almost one-quarter since 2000 (24%).

The origins of our labor market failures do expand beyond D-Tech. For example, factors completely unrelated to D-Tech may have played key roles in increasing the number of men with arrest records. In addition, as other sources of energy (notably gas) have grown, demand for coal has drastically fallen (Thompson 2016). This particular dynamic has played an important role in eliminating Jim's employment prospects in certain parts of America. Meanwhile, as the economy picks up, a modest number of these long-term unemployed are being re-employed.

Nonetheless, Jim and many other less-skilled men, and Antoinette and millions of our other youth are barely clinging to the edges of our social fabric. It will take more than just a stronger economy to allow them a fulfilling life. From the Welfare Cliff (Susie), to less well-prepared youth (Antoinette), to more unemployed men (Jim), our labor market is failing—and D-Tech *side effects* are a critical ingredient in that failure.

One last irony before we move on: horror stories abound about how Digital Technology will create a society without work (Manyika et al. 2017). People feel scared when they read such stories. Yet for millions of Americans, that nightmare is already a reality.

For those following the text of this book, we now turn from our labor market to other markets. You may also choose to skip to our discussion on <u>Risks to Our Future</u>.

OTHER MARKETS—37% OF U.S. ECONOMY (GDP)

Some Businesses in this "other" category may be succeeding rather than failing. Such industries would be those fulfilling their purposes—their reasons for existence.

We've already discussed news, social media, and software earlier in this book; those industries are failing. Moreover, *every industry I analyzed developed a fundamental failure during the D-Tech revolution—with D-Tech side effects as a critical source of that failure.* Overall, what Business delivers is not what we need.

Business is also delivering fewer products and services than it should be. We may be lacking as much as 35% of our economy because of D-Tech *side effects.*

Through productivity, economists measure how well Business is translating technology into a volume of products and services. Productivity is the measure of output per unit of input. For example, suppose we hire Daksha and Nandan to mold ceramic plates. These two employees arrive on time and immediately mold plate after plate. Over the course of the day, the two of them produce 250 plates—125 plates per person. That's a reasonable baseline level of productivity.[20] Compare 125 plates per person with what we get from Devin, Rayden, and Reuben. Reuben sits on Devin's first plate of the day. Then Rayden accidentally knocks over the potter's wheel operated by Reuben. And the day continues from there. Three boys may be working, but Devin, Rayden, and Reuben deliver only 85 plates. That delivery amounts to less than 30 plates per person! That's unproductive (Figure 5.25.)

As technology improves, we'd expect productivity to improve. A better potter's wheel may allow faster plate production, or a walkie-talkie device may allow better communication while crafting plates.

As Bill Gates, Microsoft's founder, pointed out, "Innovation is moving at a scarily fast pace (Friedman 2014)." This pace suggests we could experience scarily fast productivity growth.

Figure 5.25: The boys in the top two frames are
unproductive, using their time and capital inefficiently.

But we're not experiencing this type of growth. On the contrary, D-Tech has decimated productivity growth. Figure 5.26 shows that productivity (labeled "Total Factor Productivity") increased 1.9% per year from 1947 to 1969. In contrast, productivity has increased only 0.8% per year from 1970 to 2012.

U.S. Total Factor Business Productivity

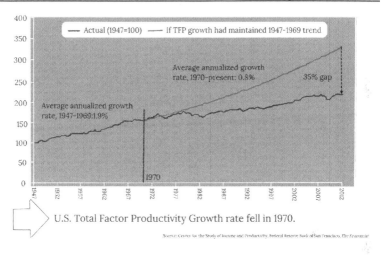

U.S. Total Factor Productivity Growth rate fell in 1970.

Source: Center for the Study of Income and Productivity, Federal Reserve Bank of San Francisco, *The Economist*

Figure 5.26: U.S. productivity growth has slowed since 1970; in other words, it appears we're benefitting less from technological development than we were before the 1970s (Sornette and Cauwels 2012, fig. 6).

Looking at the figure, we see that this slower productivity growth since 1970 has cost us 35% of our economy. If we had maintained the 1947–1969 productivity growth rate through 2012, the average American would be receiving almost $20,000 more every year.[21] D-Tech *side effects* provide the most-compelling explanation for that productivity drop-off. Over the last few pages, we've explained how these *side effects* are derailing Business's efforts to deliver what society needs. In several industries, we are finding increasingly complex markets featuring interconnected problems across Government, Culture, and Business. Business is proving inefficient at operating in that environment. In other words, in today's complex environment, Business is ineffective at delivering:

- Productivity growth.
- Product and services that satisfy our needs.

Other explanations for this falling productivity growth are less compelling than those focused on D-Tech. Perhaps the global energy crisis, for example, weakened productivity growth in the 1970s. However, we're more than 30 years past the end of that crisis.

Also possible is that the measured drop-off in productivity growth is overstated. Economists generally measure only things we pay for, but much of what D-Tech delivers is free. Facebook and YouTube, for example, are free to users and advertising is relatively low cost. Perhaps there's more produced than we're counting. In the case of the plate example, it could be that Reuben is producing 200 plates every day; he's just giving them away before we count.

However, that's not the case. As Brynjolfsson, Rock, and Syverson (2017, 8) explain, it's unlikely such mismeasurement is "the entire, or even a substantial, explanation for the slowdown. . . The same [under-counting] was undoubtedly true in earlier eras as well."

Brynjolfsson, Rock, and Syverson (2017, 8) suggest another explanation: this slowdown is a temporary phenomenon. Growth will accelerate again when Business learns to apply bleeding-edge artificial intelligence technology.[22]

That explanation for the lack productivity growth from 1970 through today is also not compelling. Long-standing D-Tech tools also revolutionized many aspects of Business. Business now comfortably applies spreadsheets, word-processors, the Internet, and cell phones, for example. Both factory and office workers' jobs have been completely revolutionized by those tools:

- Most factories have a tiny fraction of the factory workers now compared to what they had in the 1960s.
- Accountants can complete month-end exercises in hours rather than weeks.

One leadership team gave salespeople a $25,000 spreadsheet tool listing the salespeople's most and least profitable customers. Salesperson Janet and her colleagues were amazed by the insight. Customers who Janet thought were her most profitable were relatively unprofitable, and vice versa. Salespeople completely changed their focus and selling strategy. Profit increased more than 300% in two years.

Without D-Tech's *side effects*, productivity growth would have exploded with such tools. Instead, productivity growth has been crushed.

Perhaps we are failing today, but how about tomorrow? D-Tech certainly provides us the potential to design the exact society we want. What Business-related risks does society face tomorrow?

RISKS TO OUR FUTURE

Our society is at high risk. The millions of us in America (and even many beyond the U.S.) are more dependent on the decisions of a select group of individual business leaders than ever before.

Today, our winning Business winners *win*: in our D-Tech world, a winning business obtains spectacular decision-making power. In the introduction to this chapter, we highlighted how

D-Tech has driven our economy to become winner-takes-all. What's the prize?

At stake is the ability to direct large swaths of society. Every major business is responsible for incredible volumes of resources. The tone that leaders set has massive impacts on their businesses—and thus on the future of society.

Consider United Airlines' CEO Oscar Munoz. He was compelled to respond to a high-profile situation recently:

> *Blood drips down the 69-year-old Dr. Dao's chin. You just heard him scream off-camera, and now security is dragging him down the airplane aisle. Dr. Dao's glasses are dangling down his face; his bare midriff is open to the camera.*

Dr. Dao's crime? He wanted to use his airplane ticket. Dr. Dao wasn't convinced that transporting United Airlines employees was more important than getting him to his medical patients. When Dr. Dao wouldn't give up his ticket, he ended up with two lost teeth, a broken nose, and a concussion (Griffiths and Wang 2017).

Those of us outside United Airlines are unclear which of the two following responses accurately represents the true United Airlines CEO:

- **Scenario 1:** The day Dr. Dao was dragged from his plane, United Airlines CEO Oscar Munoz responded to the incident with an internal memo, "commend[ing] [the United Airlines employees] for go[ing] above and beyond to ensure we fly right (Zdanowicz and Grinberg 2017)." As Mr. Munoz highlighted, "Our employees followed established procedures for dealing with situations like this."

- **Scenario 2:** Two weeks after the incident, Oscar Munoz's response changed immeasurably. Mr. Munoz said that he ensured the company took full accountability for Dr. Dao's situation. The same day United settled the lawsuit with Dr. Dao, United released a 10-point improvement plan for overbooked flights. As Mr. Munoz explained, "This is a turning point for all of us at United, and it signals a culture shift toward becoming a better, more customer-focused airline (Loyd 2017)."

We don't know who the sincere Oscar Munoz is. Perhaps the second scenario represents the true man: he had two weeks to learn about exactly what happened, after all. Perhaps Mr. Munoz was simply poorly informed in Scenario 1. On the other hand, the first scenario could be the true Mr. Munoz as well. Millions of people around the world had witnessed the video of blood dripping down the elderly doctor's chin (Webb 2017). Public outrage was palpable. Oscar Munoz's only choice was to present his company as if it were sorry for the incident.

The answer matters. The future of our airline travel will be determined by the decisions Mr. Munoz and other business leaders make. Will our seats squish us even closer together in the future than they do today? Will airline personnel show gracious respect to the exhausted Mommy trying to get little Rayden and Reuben home to Gramma? And so on.

For most of these decisions, the voices of millions of people won't force the hand of a CEO in one particular direction. If the true Oscar Munoz is shown in Scenario 1, then he is coaching his entire company to focus solely on efficiency. Get those folks up into the air, and get them down. If you need to break a few ribs and teeth, then you are simply "going above and beyond to ensure we fly right." Just make sure you don't get caught on camera next time. Those following Oscar Munoz's lead will focus strictly on efficiency, ignoring humanity.

If Scenario 2 shows the true Oscar Munoz, on the other hand, then humanity could triumph. United Airlines employees would build a business that creates profit by being humane (and then getting paid for that).

Many of these winning businesses have greater impact on society than previous business winners. During the 1920s, for example, a handful of business magnates dominated the U.S. economy (Smiley 2004). These magnates unquestionably could spend their money to impact how society operated. Yet, these 1920s magnates were limited in the directions in which they could apply their business resources. A steel company, for example, would make steel. A car manufacturer would manufacture cars.

Today, D-Tech magnates play critical and powerful roles in determining how we think. These winners also play massive roles in determining society's future, for example (Epstein 2016):

> *Reddit has frequently been accused of banning postings on specific topics, and a recent report suggests that Facebook has been deleting conservative news stories from its News Feed, a practice that might have a significant effect on public opinion—even on voting . . .*
>
> *When Google's employees or algorithms decide to block our access to information about a news item, political candidate, or business, opinions and votes can shift, reputations can be ruined, and businesses can crash and burn. Because online censorship is entirely unregulated at the moment, victims have little or no recourse when they have been harmed.*

Google was recently sued by an organization claiming that Google should not have the right to censor its own website—because the public interest is too large (Gardner 2017b). Google responded, "Giving viewers the choice to opt in to a more restricted experience is not censorship (Gardner 2017a)." Whatever the legal truth, the question remains valid: what should Google be allowed to do? Whatever it does will massively impact society.

What type of impact does Google have? Recently, Google determined a drug-treatment, commercial policy for the entire world. Temporarily banning advertisements for addiction and rehabilitation centers eliminated a popular means for people searching for treatment options.

The impact of this one corporate decision is breathtaking. Clients were deprived of a highly visible, and perhaps highly material method for reaching addiction centers—based on a private decision.

Google did have a constructive reason for these particular bans (Coldewey 2018b):

> *[Google] was acting as a platform for shady referral services earning huge undisclosed commissions. . . Essentially these ads, which commanded huge prices on Google's networks, would show for people seeking addiction treatment; the help lines and services listed would then refer the person to an addiction center. These centers were, unknown to their new patients, paying enormous finders fees to the referral services, on the order of tens of thousands of dollars.*

Nonetheless, the impact of this one corporate decision was astounding.

Public outrage at Facebook recently piqued, as well. A particular photo of naked children had become fairly common on the social networking site. The outrage wasn't, however, that Facebook was allowing child pornography. The image Facebook banned was the Pulitzer-Prize-winning image of children (one naked) fleeing a U.S. napalm attack during the Vietnam War (Singer 2016). That image plays a powerful role determining society's view of American participation in that war. Right or wrong, a single, private company is playing a particularly powerful role in determining society's view of history.

As D-Tech continues to evolve, the impact of these few winners will expand, as well. Consider the impact of technology on the impact of human decisions. When I feel frustrated, sometimes I like to skip pebbles across a lake. I feel better afterwards and no harm is done.

In contrast, today, Elon Musk and many others are working to integrate the human brain with D-Tech. We can already control a computer with our brain waves through a small hat we put on our head (Kurzweil 2018). That connection enhances the strength of our thoughts. What happens when a frustrated person decides to skip our moon across the solar system rather than pebbles across a pond? As the power of technology grows, actions that are superficially similar have different implications. And with different implications, society's design will again matter in new ways.

D-Tech has driven us toward a winner-takes-all economy. This technology has confused our Government and Culture; these subsystems of society are no longer leading society. Business leads. And a few men (even fewer women) lead these winners.

Winners do retain their throne for a shorter period of time than ever before (Anthony et al. 2018), yet such statistics can be misleading. Small competitors using the same technology are unlikely to take down the current winners through adept tactics. Today's winners are overthrown when new waves of technology supplant the current generation; the winning companies are frequently replaced, along with their tech.

Nonetheless, business' winners lead society for as long as they continue to apply the most appropriate technologies. The decisions such leaders make will determine the future of our society—at least until the next generation of technology takes these leaders down, supplanting them with our new leaders.

We don't get to choose these people. Investors do. We can only be aware of these people's powers, help them understand their roles, and engage.

And, if you are one of these people, you must recognize your out-sized responsibility.

SUMMARY: IMPACT OF DIGITAL TECHNOLOGY ON BUSINESS AND OUR ECONOMY

Digital Technology has transformed Business. *Directly*, Business benefits from D-Tech. We can instantly know the status of our global supply chain shipments around the world. And it's easy to revise marketing materials dozens of times.

However, D-Tech also creates *side effects*, or emergent impacts that are making Business's outcomes less beneficial to society. Change is becoming faster and our ability to change our environment is expanding. When appropriately targeted, the effect is positive. However, when that change fits society's needs poorly, the damage is also greater. Today, we're increasingly targeting product and service labels that poorly reflect what we actually need.

While we need to target our technology more precisely, we're also zooming along a massive, three-dimensional highway: we can move in any direction, accelerate or slow down, and more. There are few lane markers yet everyone must avoid everyone else. Our decisions are deeply interconnected. A decision to brake by someone hundreds of meters ahead of us can lead to a chain reaction of braking, swerving, and eventually colliding. Chain reactions like these travel through IT systems, parts, products, regulations, and more. We must collaborate.

We have not successfully redesigned Business to manage these new challenges the way successful leaders of the Agricultural

Revolution did. We have not benefited from an equivalent to the Qin Dynasty's first emperor who re-enforced the role of money and standardized the size of carts, weights and measures.[23]

Rather than rising to these D-Tech challenges, each type of Business's action is now compromised in critical ways. In key areas, Business is focusing on creating the most attractively labeled products. As D-Tech allows us to shape our world more profoundly, we increasingly create superficially wonderful products whose unlabeled aspects are allowed to wither and become sickening.

Execution is increasingly overwhelming. Constructive decisions require that rising numbers of people, machines, and systems collaborating effectively. The results of ineffective coordination are growing: destructive consequences, failed collaborations, and major issues "hiding in plain sight."

On the horizon are key decisions about how to influence customers with D-Tech. Undoubtedly, much of this influence will be targeted constructively. Unfortunately, Business's record of influencing us in the past (with traditional advertising) appears more destructive than constructive.

Business is currently failing society. Over 60% of our economy is dedicated to industries that are failing to deliver what society needs from them. Productivity growth has fallen. Our future is more and more dependent on the decisions of a few Business leaders.

Our Business leaders are our new royalty. However, their games are more complex than anything that existed in medieval times. Despite all of this, it would be foolhardy to think we could do without Business. Our only recourse is to align the challenges of Business with those of society. We need our knights in shining armor. We need the pageantry of success that comes from noble warriors slaying villains. However, those slain must be society's villains, not society's heroes.

In many respects, we are unlike the successful societies from the Agricultural Revolution. We are not progressively evolving our Business institutions in the face of our challenges. Through our rapidly improving D-Tech and other technologies, we are sowing the seeds of an amazing future, complete with both physical and virtual sides. Thus far, however, we are not updating and upgrading the Business structures so they will nurture such efforts. Our society is drowning in potential, and Business is not doing enough to support it.

In our next and final chapter, we look for solutions to our problems. If our Culture, Government, and Business are breaking down in the face of our D-Tech environment, then we must remake them. Can we strengthen our systems to operate under the increasing stresses of D-Tech?

The risks for society are high. In our final chapter, we find our solutions.

CHAPTER 6

SOLUTIONS:

Vision Beyond the Bottom Line and Other Remedies

How do we solve society's problems?

We follow much the same discovery process we did about the Agricultural Revolution, the focus of Chapter 1.

During the Agricultural Revolution, specialization developed. The key challenge was to develop a society allowing such specialists (initially largely in farming) to be effective. Successful societies introduced the Business / Economy, Government, and Culture tools needed to facilitate collaboration among specialists.

We've now spent 10,000 years developing the tools that help specialists prosper. We're skilled at applying such tools. And, with D-Tech, our specialists can accelerate creating new insights—and even more new types of specialists. What motion does a bat's wing take as the animal flies through the air? What are the molecular patterns by which proteins fold? How can we land a human on the planet Mars? We are beginning to master such questions.

Yet, society has come full circle. We are now more challenged to operate in areas where we are *not* specialists. We can find specialist information, such as on the motion of a bat's wings—but find it more challenging to bring such information together for effective decision-making. Our approach must change. We must be able to bring together all of the insights created by our specialists—in a way that makes our day-to-day lives and decisions more nourishing and fruitful. In business, in particular, we must develop vision beyond the bottom line so we can find the most value-creating approaches to collaboration.

How do we find the best way to collaborate in today's complex world?

By considering the problems with my old car, we see the keys to diagnosing our current problems. Let me explain:

For a number of years, I had a Honda. It was dependable, cheap to maintain, and . . . *boring.*

Things have gone well for me recently, so I wanted to switch to a Ferrari. Things haven't gone well enough for me to afford a whole Ferrari, so I decided to upgrade my Honda instead.

I hired a consultant to benchmark my Honda against the Ferrari. The consultant declared, "Your Honda is slower than the Ferrari."

And I said, "Yes, I know."

Then the consultant said, "Your Honda is slower because your Honda's engine is weaker than the Ferrari's engine."

And I said, "Really? Wow! I hadn't realized that." I thought that insight was enough for me to upgrade my Honda.

Now, I knew full well that relying solely on benchmarking for strategic planning in a complex society is dangerous, but I had concrete *facts.*

So, what did I do? I bought a Ferrari engine and I installed it in my Honda. My *Ferronda was* unique—yet it worked poorly.

The body of my new car was constantly shaking; the shock absorbers in my Honda aligned poorly with the Ferrari engine.

My *Ferronda reeked!* The constant black fumes discharged from my *Ferronda* were overwhelming. The reason? My Honda exhaust system worked poorly with the Ferrari engine.

My original dream was, "I'm flooring my *Ferronda* out of my driveway as I drive my kids to school." But the tires on my Honda weren't built to grip the road when powered by a Ferrari engine. And even if they had been, I couldn't have controlled this beast with my Honda steering.

As I watched the little kid in the go-kart pass me by, I sighed, and wondered what I had done wrong.

Like much of society today, my *Ferronda* is a poorly working system. Individual pieces are highly impressive—the result of incredible D-Tech analysis. Yet, those individually impressive parts collaborate poorly.

We can think of our broken societal systems in terms of the *Ferronda*. Consider our educational system. Our teachers are the Ferrari engine: they are well-educated, most care deeply about our children, and they are even provided ongoing training. Our creative programs to take children who are behind grade-level and catch them up are a set of Volvo shock absorbers: considering solely such children, the teacher's time focused on these children is well spent. The system designed for physically and emotionally challenged children to mainstream them as much as possible are the Honda structure: designed to keep everything together.

Each of these parts, looked at individually, is exceptional. Yet, in combination, they are a disaster. Why? We continually try to upgrade one part, or another part, or another part—rather than integrating the system.

Figure 6.1: Combining a wonderful Ferrari engine in a previously-dependable Honda fails. Any time you combine elements that are individually wonderful but collaborate poorly you get such a mediocre result.

Which societal systems are operating as *Ferrondas*? Any societal system in which the outcomes are unaligned with the system's purpose (which is related to mission and vision) is a Ferronda. If the outcomes of the system are unaligned with the system's purpose, and are not on track to achieve that purpose, then we are facing Ferronda-type, systemic issues.[1]

Over the course of the previous five chapters, we've highlighted three classes of issues:

1. **Gathering Information:** We obtain exceptional detailed information. The tailpipe of our *Ferronda* is vibrating 19 times per second. Fourteen grams of gasoline is flowing to the engine every tenth-of-a-second. In areas where we specialize, such information is enlightening. However, we lack the structure (the ontology) to bring that information up to a level of generality understandable by non-specialists, such as "Your *Ferronda* needs a tune-up . . . *NOW!*"

2. **Understanding Context:** We need to interpret the information we receive in order to understand how our world operates. What does the information we receive imply for our broader world context? We may see some red lights on the dashboard. We see the mountains ahead with a white substance at the top. We need an insightful voice telling us, "The *Ferronda* was designed for city driving. You are about to enter some icy mountains that require completely different capabilities. You'll either need to re-design the vehicle or turn back." Humans provide much greater capability to deliver such insight today.

3. **Analyzing Options:** We need to collaborate more deeply than ever before. No part of the *Ferronda* delivers anything meaningful on its own. Only through effective collaborating across all 20,000 parts do we get transportation combined

with a targeted human experience. Today, our society requires a similar level of coordination: we need broader collaboration, across more people, than ever before.

Provided this perspective of the *Ferronda*, and the identification of our three core issues, we can now identify solutions.

HOW DO WE FIND WHAT WE NEED? MARKETS FOR PRECISE CONCEPTS

We need data summarized in a way that has meaning for us. Detailed metrics and information is available about our *Ferrondas*, such as "Your engine is currently 348 degrees Fahrenheit, you have 0.38 liters of oil with a viscosity of 3.8 centistokes, and your exhaust emissions have an average size of 76 nanometers."

Such detailed information is equivalent to the information we easily find about our food. We can find the details about our chicken pot pie's saturated fat levels, its likely sugar content, and the relative vitamins in the peas. Yet, what we want to know is whether or not the combined food is healthful—but we struggle to get such a summary.

In our cars, if we need to *pull over*, a warning light shines on the dashboard. If we can get such a straightforward light on our dashboard, why can't we get similarly easy-to-consume information about our chicken pot pie? Can we get such easily consumed information flexibly delivered?

Gathering information efficiently means structuring and combining the information that is so readily available today. An automotive engineer designed a single equation that combined engine temperature, oil viscosity, and emissions—to indicate when a driver should receive a message to pull over.

Information theorists call such information structures onto-logies—definitions of the meaning and relationship of various elements of information. An example of a simple ontology is: living things are divided into the categories plants, animals, fungi, and bacteria. From that structure (the ontology), you know that the total number of living things is the sum of the things in each of those categories.

Yet, the ontology, by itself, is not enough. Our knowledge continues to change rapidly. Biologists, for example, recently realized that we have two different types of bacteria: Archaebacteria and Eubacteria. An updated ontology would thus say that living things are divided into the categories: plants, animals, fungi, and the two types of bacteria. We need flexibility. So, how does society structure itself to flexibly develop and apply specialized information? Since the Agricultural Revolution, markets have been the key to effectively applying specialist information.

In today's rapidly changing world, we need flexible onto-logies—based on markets—because the information available to us changes. That flexibility requires we integrate ontologies with society's sub-systems, particularly Business/our economy. In the terminology of social scientists combined with logic, we need to apply the tools of society and mathematics to create flexible ontologies. We need precise concepts shared and developed through a market.

Let's first consider how markets work.

THE POWER OF MARKETS

We humans have masses of knowledge. Yet, we don't want to *know* all of that knowledge. We just want the benefits of that knowledge in terms of higher-quality decision-making. For such "invisible insight," we must apply market power.

A young engineer, Akwete, illustrates market power:

When Akwete enters the showroom in a furniture store, she walks up to the closest sofa to the door. She leans over and begins running her fingertips down the seams in the fabric along the edges. Her head is cocked to one side as she feels for the precise placement of the staples.

It's only after Akwete is confident she knows exactly where the staples are in the sofa that she sits on the sofa. Upon sitting, she acts in many ways like I do at a sofa showroom. She bounces up and down a bit. She leans against the backrest, and she watches as the fabric and cushioning react to her weight.

You see, Akwete is a young engineer. Her role for her employer, a major sofa manufacturer, is to redesign their sofas' stapling. Her manager estimates Akwete can reduce the cost of the average sofa by almost 2% just by re-thinking stapling. Akwete will spend 850 hours over the next 8 months poring over calculations, blueprints, and staple guns.

Akwete's employer will save millions of dollars a year. Over time, competing manufacturers will learn the keys to Akwete's insights, integrating the cost advantage Akwete created. The price of sofas will be competed down.

We will all benefit from Akwete's efforts. Yet, we don't really want to know the specific results of her findings.

We want to sit in the sofa in the showroom, be comfortable, be attracted to the sofa's look, and be positively surprised by the reasonable price.

Dozens, if not hundreds, of engineers likely designed the sofa in your living room. Akwete designed the stapling, Kasem upgraded the springs, and Geni designed the cushioning. None of these people know each other, yet all of their insights are present in the same piece of furniture.

The key is the market. By purchasing a completed spring, the sofa's assembler was able to benefit from Kasem's insights—without understanding how to engineer a spring at all. Hundreds of specialists. Dozens of parts. Each sold as a self-contained sofa element. All to make you comfortable—without you having to know anything about how your comfort was designed into the magical sofa. That is the magic of the market.

So what is the unique market challenge for our D-Tech world?

TODAY'S PRECISE CONCEPT MARKET CHALLENGE

As we discussed throughout the book, through D-Tech, we have been feeding our specialists. Any factoid about our world is trivial to find. How many proteins can the human body generate? Approximately 20,000 non-modified (canonical) proteins and millions of modified proteins (Ponomarenko et al. 2016).

Yet, we frequently do not want all that detail. We want answers that would *combine* such technical detail. We don't want to know the latest finding about how antioxidant levels impact our chance of getting cancers. We *do* want to know whether apple A or apple B is more healthful (as a result of their different antioxidant levels).

Our challenge is to make consuming information easier. We want to consume combinations of information as simply as we consume a sofa, which is made of a combination of parts. In other words, we want experiences like the following:

> *In the supermarket, Jennifer takes a picture of two fruits and then tells her friend Julie, "Wow! According to the application Apple Health, this Granny Smith apple is almost twice as healthful as that Pink Lady apple!"*[2]

Julie responds, "Wait a minute. I'm diabetic. Let me check the app, Diabetic's Fruit Check. Wow! The answer there is even more strongly in favor of the Granny Smith."

A few days later, Dr. Stracten publishes a journal article with new findings on the health impact of antioxidants. Without anyone at the organizations creating the healthfulness applications doing anything, the answer in both applications changes—shifting roughly 18% in favor of the Pink Lady.

And then the fiasco occurs. Dr. Stracten's definition of anti-oxidants is torn apart by his academic peers. Leaders across the nutrition industry switch from Dr. Stracten's definition of which molecules should be classified as antioxidants to a competitor's. The creators of both Apple Health and Diabetic's Fruit Check shift their definitions along with everyone else. As a result, Granny Smith apples healthfulness gains roughly 5% relative to Pink Lady.

Meanwhile, based on such analysis, the following conversation occurs within Apple Orchards, Inc., America's largest apple grower. Managing Director Jim asks Dr. Figueroa, the head of research, " Juanita, our apples' low antioxidant levels are killing us financially—and our clients physically. How can we improve them?"

Juanita smiles. "Do you mean for this next growing season or do you mean longer term, say 5-10 years from now? My team has ways to impact both. Short term, we're probably talking about fertilizers. Longer term, we can talk about breeding, genetic modification, fertilizers, soil enrichment programs, processing options, and more. I could have a brainstorming session with my team and I'll bet they'd come up with a half dozen more approaches."

Juanita laughs, "Jim, you know this is the first time you've asked me how to improve the healthfulness of our products? We've been focused on cutting cost, improving taste, and increasing glamour so long that we'll have to pull out some of our old textbooks to help us."

Such conversations contrast with those described over the course of this book. These conversations are based on our specialists' most up-to-date knowledge. Yet, the information is easy-to-digest. Such information access would be empowering—and it could eliminate the impact of many information-gathering problems described over the past few chapters.

Despite being so radical, access to such information also parallels existing, mundane activities such as buying a sofa. The health score itself is the completed sofa. Just as a designer or engineer defines a particular sofa, a qualified nutritionist could define how to score the healthfulness of an apple. Presumably, apple healthfulness is based on levels of antioxidants, sugars, volume of roughage, and the like.

Meanwhile, the elements of such an equation are like the parts in the sofa. Perhaps the molecular definition of an anti-oxidant is equivalent to a sofa's spring: critical to the outcome and highly technical in its design. One definition of antioxidant could easily replace another—just as one spring replaces another in a sofa.

What's the key? Creating the appropriate market for structured information—to mirror our current market for physical products and services:

- Society must develop the standards that will allow the market to thrive.
- The language in which the concepts are created must be Digital Technology's language—mathematics.

STANDARDS FOR MARKET DEVELOPMENT

Why is the market critical in order to develop ontologies? Without a market, designing a meaningful ontology is equivalent to my designing a sofa from raw materials such as cotton balls and iron ore—all while sitting in my living room. I'd have to design the springs, and cushioning, and stapling—all myself. *I don't know how to design a spring, and I certainly don't know how to smelt iron ore!* I'm sure I could find information about defining a spring on the Internet—but that would be insufficient. We need a market in which an organization specializing in springs designs the springs. An organization specializing in cushioning designs the cushioning, and so on.

From today's perspective, it may seem like markets simply spring up. Every nation around the world today has developed markets in which businesses sell their wares.

Yet, that is not so. Meaningful trade blossoms only once standards for markets are defined and an overwhelming reason for commerce exists. Rules are frequently the earliest standards. As discussed earlier, the Code of Hammurabi is known for clarifying rules and setting minimum standards for conduct in Babylonia four thousand years ago ("Code of Hammurabi" ca. 1780 B.C.E.).

In China, Emperor Qin Shi Huang standardized units of weight, length, and currency ("Qin Shi Huang" 2014). I will pay you $4 for 2 kilograms of potatoes is simpler to compare with other offers than, "I will provide you a 15-minute massage for that pile of potatoes."

In Japan, commerce thrived as an overwhelming reason for commerce was instituted: governors from around the country were forced to spend half their time in their own states, and half their time in the national capitol. The wealthy governors had to find ways to provide the goods that would sustain them in both locations (Ito 1991).

We can directly apply this learning to creating markets for Precise Concepts. We must define what is acceptable and unacceptable in such markets. Poisons, for example, must not be considered a nutritious element for fruit. Government has an important role in defining the principles determining such limits.

A sponsor recognizing the commercial benefit must define the markets' standards, the equivalents of the weights and measures. It may not matter, for example, whether we divide inches into halves or thirds. Yet, if we use different standards, we will have problems: my measure of 1/3 inch cannot be precisely translated as 0 or 1/2 inch according to your standard.

And, that sponsor must recognize the overwhelming commercial drive for the market. Precise Concepts provide us flexible ontologies for combining today's detailed knowledge. Anywhere the flexible application of such knowledge could materially improve decision-making is likely to have an overwhelming commercial drive for the development of an active market. For consumers, information related to diseases and healthcare, for example, jump out.

For business-to-business applications, if design, architecture, and engineering play important roles, the application is promising for a flourishing market for concepts. Examples include military weaponry design, electronics design, and building architecture.

A market for such ontologies, meanwhile, allows all of our specialists to continue specializing in their areas of expertise. If you are an expert in designing springs, design the portion of the ontology associated with spring design. Sell your completed concept to the expert who is focused on designing total sofas.

What happens when a single organization attempts to define an ontology? In chapter 5, we introduced the ICD-10 codes for healthcare. American physicians must use this list of 69,823 codes to diagnose their patients today. This list—an ontology—is weighed down by such obtuse diagnoses as, "being sucked into a jet engine," "being struck by a duck," and "being burned when water skis catch fire" (AAPC, n.d.).

Individual organizations attempting to define a complex ontology by itself, without such an information market, are bound to end up with such a mess. Virtually every company (90%) is expected to hire a Chief Data Officer (CDO) by the end of 2019. The role of the CDO is essentially to ensure that the appropriate ontology is available to the organization—and to ensure the corresponding data is relatively clean. Only *half* of CDOs are expected to be successful by the end of 2020 (Gartner 2016). A vibrant market for precise concepts could go a long way towards improving the chance of CDO success.

What happens when a market for ontologies (precise concepts) is created? We are able to flexibly apply the masses of information already being created. We are able to deliver such information in a manner straightforward for the end user/consumer. We redefine our society's ability to leverage our knowledge for good.

THE LANGUAGE OF PRECISE CONCEPTS

The language of precise concepts is mathematics (or, more precisely, logic). A mathematician would suggest the solution must be "semantically precise ontologies" (Omohundro 2015).[3]

What is the challenge we face? We are dependent on D-Tech to find us the information we need. Thus, the concepts we create must be understandable by D-Tech—i.e., they must be in a mathematically precise language.

With mathematical precision, computers can do the work of combining and structuring the information we need. The machines can provide us easily consumed information—that still incorporates all of the up-to-date, detailed knowledge available. Computer scientists have described some of the benefits of such precise concepts:

- **Buildability:** We want a tool where the definition of antioxidants in measuring the healthfulness of an apple is as easily replaced as the springs in a sofa. With semantic precision, that replacement is as straightforward as replacing 'X = 8' with 'X = 10' in a grade schoolers' algebra homework—and the computer can execute that switch. In addition, we want the masses of information currently available on the Internet and in private data stores to be, at least in large part, automatically incorporated (increasingly, over time).

- **Expandability:** A concept developed by one person can be purchased and used somewhere else. Once Apple Health has defined a metric for the healthfulness of an apple, for example, the Orchard industry can use that definition. Perhaps the other party creates an automated set of standards by which all "certified" apples will be sold, including healthfulness and standards for care of the environment. All standards that apply such an Apple Health definition could be automatically updated by D-Tech any time Apple Health is updated.

- **Consistency:** The consistency of our rules, regulations, and standards can be automatically checked.[4] We face more than 250 million federally enforced legal constraints on what we can and cannot do. If we have semantically precise definitions of those rules and regulations, the computer can identify exactly which sets directly conflict with each other. The computer can also automatically check whether the technical specifications for our new power plant meet all of the environmental regulations—with the click of a button!

- **Verifiability:** We can insure, with absolute certainty, that any and all specifications meet desired criteria. As discussed in Chapter 2, the impact of mistakes is growing. For a self-improving "general artificial intelligence," even a seemingly trivial oversight could be massively destructive (Russell 2017).

In short, reducing the degree to which we are overwhelmed by information requires a market for precise concepts. Today, in our information world, it is as if we have to purchase every element of our sofa individually. I don't want to have to go shopping for sofa springs! We need a market for concepts that will allow us to purchase the easily consumed information equivalent to a fully formed sofa.

To allow D-Tech to develop such information, these concepts must be defined in the language of digital technology—mathematics. Others are calling for exploration of such semantically precise concepts (see, for example, Marcus and Davis 2018). Yet, today such voices are in the minority: it is business that must begin demanding such exploration for this mathematically precise approach to flourish.

HOW CAN WE ENHANCE UNDERSTANDING? REPUTATION-BASED STORYTELLING

We need to understand how our world works. Before we decide whether we want a Honda or a Ferrari, we need to understand whether our world is more conducive to one or the other. Is our life entering us into a road race in which the Ferrari's speed and handling will be valuable? Or are we about to have children for whom we will more highly value the utilitarian Honda?

Today, we receive all the information about our world in bits and pieces. We receive *pings* and nudges. We're pushed and pulled into thinking our world is about to act in a particular fashion. Yet, what is our world truly like? How can we understand our world?

Understanding the world broadly is a human endeavor. Since the beginning of human history, we humans have developed such understanding through storytelling. Our brain has been developing for millennia to solve storytelling challenges in which both content and context are critical. The confusion we face across society—the challenges understanding our business, government, and cultural situations—are all storytelling challenges.

Today, we must recognize the historical elements of the storytelling challenge. We must also take the challenge one step further, so that we can support constructive storytelling in our more complex D-Tech world.

For millennia, humans have earned sterling reputations for telling stories that promote informed decision-making. Once again, we must leverage that skill.

STORYTELLING INTRODUCTION

Digital Technology may give us loads of facts, but we humans make up the story. We create sophisticated stories about how our world works given even the most basic information. The more fully the information we are given fills in the blanks for us, the better we understand the implications.

What is the difference between a set of facts and the story we create about those facts? A friend of mine, Bob, helps illustrate:

> *When Bob was in the military, his superiors would tell him, "When you're reporting to me, son, just give me the facts. Don't give me any of your stories."*
>
> *And so, what would Bob say? He'd say, "Sir, between 2013 and 2017, 133 American servicemen died in accidents in air vehicles. That's a 40 percent increase from the period before 2013."*

Such facts invite the audience to imagine their own context. How is the senior officer likely to respond? "Son, we're in the business of war. We've had some amazing aircraft delivered to us recently. As we test those vehicles to their limits so we can use them to win that next war, some people die. Those deaths are regrettable. But, it's the cost of war."

The lieutenant understood the facts in the report—but made up his own context. To the lieutenant, the world was amazing machines and uncoordinated humans. The data represented images of sophisticated airplane test flights and experiments pushing helicopters to the edge of their abilities.

Reality is different from what the Lieutenant's imagination portrays. Rather than leaving context to the listener, we need to tell a story in which we provide the reader clear context. Through storytelling, we're likely to align when it comes to the truth.

What is actually happening today? Bob's report communicating that truth would go something like this: "Sir, I went to the hangar to requisition a helicopter for our training tomorrow. On paper, we have six helicopters. In reality, we have two working helicopters and four scavenged for parts spread across the hangar."

The young serviceman would continue, "We used to have six senior sergeants with gray hair. Now we have one. And when I asked the gray-haired Sergeant Louis for the helicopter for training, he shook his head. He said he would do his best. However, getting a working helicopter for training is increasingly hit and miss."

"The reality is," the young officer would report, "In 2011, the U.S. government changed the way that it requisitions military funding. The government shortened the planning cycle from several years at a time to individual year budgets. At the same time, they cut military spending. We couldn't cut long-term projects immediately. So, what did we cut? We retired the

> *senior officers with experience while shaving repair and main-tenance budgets (Copp 2018).*
>
> *"What's the result of this fiasco, sir?" Bob would conclude, "The rising deaths of American servicemen are not the cost of war. These deaths are the cost of bad planning."*

Storytelling provides us the context surrounding the facts they present—it "fills in the blanks." In other words, stories provide the deep insight about how our world works that facts alone cannot.

The key to supporting constructive storytelling in our D-Tech world is the topic of the rest of this section.

TODAY'S STORYTELLING CHALLENGE

As we discussed throughout the book, through D-Tech, we have exceptional ability to suggest context. On the one hand, D-Tech provides powerful ability to manipulate us (using sug-gestion)towards a particular, desired outcome—constructively or destructively.For example, all too frequently, the *Ping!* of our phones diverts our attention from those in our vicinity. At the same time, D-Tech, itself, cannot understand the complex world in which we live—technology is only a tool, used under the guidance of a human master. Only humans understand broad context. The face of our challenge today is Bernie Sanders . . . in a muscle pose (see Figure 6.2).

Agents working for the Russian government posted this ima-ge—from a real coloring book (Daddona 2016)—on Facebook, on behalf of a (nonexistent) lesbian, gay, bisexual, and trans-sexual rights group, "LGBT United." The post invites Americans to "stop taking this thing [the U.S. presidential election] too serious" and instead enjoy the coloring book "full of very attractive doodles of Bernie Sanders in muscle poses."

Through this post, the Russians were telling the American LGBT community a story: the American presidential race of 2016 is frivolous. With such a perspective, fewer in the LGBT community might bother voting (Ring 2017).

And how many resources were expended on this endeavor? The Russians spent roughly $100,000 on all of their social media efforts—which amounts to no more than one five-thousandth of one percent of the Russian federal government budget.[5] Nine million Americans have a salary of at least that amount. That spending by the Russian government, which spends billions of dollars, is equivalent to someone earning $100,000 per year spending ten cents on such content.

How do I look at this occurrence? A human, posing as someone who didn't even exist (much less have a strong reputation), used limited resources to tell a story—with bad intent. The facts were all true—the coloring book exists—yet the intent was to give a false perception of the world around us: that a presidential election doesn't matter.

Our challenge is to support constructive storytelling rather than the use of muscle-posing Bernie to discourage voting. Yet, machines do not understand our world clearly enough to tell a coherent story. *Technology doesn't solve problems—people solve problems.* We can support constructive storytelling by focusing on the *human*, rather than the machine; that is, by:

1. Strengthening our system of reputation, which rewards humans for constructive storytelling.
2. Increasing awareness that our communication, including D-Tech sourced communication, is in fact storytelling.
3. Continuing to expand our efforts to ensure humans teach machines to tell more constructive stories.

Figure 6.2: A post by the imaginary Facebook group, LGBT United. The post suggests people play with the Bernie Sanders in muscle poses coloring book rather than "take the election so serious." (Image from House Permanent Select Committee on Intelligence: https://democrats-intelligence.house.gov/uploadedfiles/6039834537179.pdf.)

STRENGTHENING REPUTATION

First, *Reputation* is evaluation based on the recording of an individual's or group's actions and stories over time. For example, Bob's insightful story about the state of American military aircraft should strengthen his reputation—that story helps his audience understand the context of the data presented, it challenges their worldview, and it facilitates better decision-making. As individuals earn a strong reputation, they are rewarded with greater influence, respect, and opportunities.

As a society, we must reward such constructive storytelling because developing the story's insights requires time and effort. Consider, for example, the challenges delivering insightful stories about American culture. Such stories rely on understanding perspectives across American society.

A person learns about different cultures through cross-cultural relationships. Teenagers developing an understanding of cultures other than their own, for example, must experience life outside their comfort zones. Such Americans must have friends who are black, white, Asian, and more. They must explore what life is like as a Guatemalan in America, or as an African American. As teenagers, such learning teens will make jokes with their friends as part of their learning. And, because those jokes are based on respect and consideration, most will be in good taste.

Yet, sometimes teenagers go too far. Do you know what happens when your inappropriate joke hurts a friend? You feel that friend's pain. You get a deeper understanding of what it's like to really be them. You understand what it is about that joke that makes it inappropriate. In short, pushing yourself to the edge, and sometimes over the edge, provides you the insight that helps you deliver incredible storytelling.

When a reputation system works well, such pushing of yourself pays off. You develop insight over time. Sometimes you make mistakes, but such mistakes teach you. The wisdom of your stories, and your reputation, improve over time.

Today, too frequently, we do not consider reputation. Instead, we focus on solitary events. Consider the recent case of the white teenager, Noah, who asked a date to his prom (a promposal) by posting a picture of himself holding a poster saying, "If I was black, I'd be picking cotton, but I'm white so I'm picking U 4 prom" (Rosenberg 2018).

That racist statement is blatantly wrong and has no place in modern society. The teen who posted that statement should be punished.

Yet, that one incident, alone, should not define Noah's reputation. It's possible that Noah has a bunch of friends who are black, white, Asian, and Hispanic. It's possible Noah is trying to understand what it's like to be black in America. Noah did something *very* wrong, yet it is possible Noah's goal is to become an insightful leader.

With a weak reputational system, however, Noah's future is now sunk. The majority of internet information about Noah covers this one horrible incident.

I don't know Noah. Perhaps Noah is an ass, belittling others unlike him as frequently as possible. Yet, by destroying someone's reputation based on a single incident, the message to others is—never challenge yourself. Never learn by going to the edge of society and filling up with insight. Never develop the perspective with which you can become the best storyteller—upon which the rest of society can rely on for wisdom.

A stronger system of reputation requires efforts across society's sub-systems:

Culture must precisely define what we want out of storytelling. For example, we need clarity—as a society—about the types of storytelling that we will support:

- Which principle, if any, is broken when I build software that spews positive sentiments across the Internet about my preferred Senatorial candidate?
- What values, if any, are trampled if I automate advertisements that nudge depressed diabetics into begging doctors to prescribe limited-benefit pharmaceuticals?
- What societal rules, if any, are undermined when I support a set of hateful posts by participants in a terrorist organization (Fingas 2018)?

As discussed previously, there are a handful of efforts to begin developing such clarity. Yet, these efforts remain limited relative to the scope of what we must define. We need to invigorate and integrate such efforts.

Government must define the limits of acceptable storytelling. "Currently, the 1996 Communications Decency Act offers near-comprehensive immunity for [digital] platforms" (Lazer et al. 2018). As we discussed in Chapter 4, this regulation frees businesses to twist perception of reality freely, including in ways that are equivalent to blatant lying. We discussed the example of Yelp, which is free to present solely negative reviews of a primarily positively reviewed business. Immunity for such deceit cannot stand.

Business will likely determine who will record reputation. Today, organizations such as the Better Business Bureau (BBB) record the reputation for businesses. Credit bureaus develop a reputational score for a consumers' credit worthiness.

Which technologies and what business models will best support recording reputation? That question is beyond the scope of this book. Perhaps the most attractive option centers around police-style dashboard cams recording everything our politicians say and do. There are likely many types of people willing to pay for such recording for reputational purposes. While made half in jest, such a concept certainly illustrates reputation's power.

REINFORCING AWARENESS

Second, to support constructive storytelling we must recognize stories for what they are. We need to recognize our own tendencies to imagine a context for the data we receive. As Jill Nephew (2016) explains, data is not "truth distilled into a more pure and palatable representation." Data is the grist for storytelling—and nothing more.

Storytelling can be energizing. In the early evening, I'll lie in bed telling stories to my children. The stories have limited content, yet they are met with laughter and giggles. I use fewer than 125 words to tell the story of Sam the Goat, but my sons imagine a farmscape stretching far beyond those words. There's an old rickety barn, a goat debating between good and evil, and an old horse that acts like a comedian.

Each child adds unique color and detail to the mindscape. Yet, by clearly providing a starting point for their imaginations, I send them off on their shared journey.

Can you imagine Sam the Goat's barn? We adults are the same as our children, although we may not admit it.

Business is beginning to recognize the power of storytelling. The CEO of Amazon, Jeff Bezos, requires boardroom communication be narrative (Gallo 2018). Any pretense of "content only" is removed; bulleted slides are banned. The term "narrative" may sound more serious than "storytelling," yet the meanings are closely related.

When listening to others, we must transform into "context hunters." Even seemingly neutral facts presented on a web page are, in fact, setting context. Text, fonts, colors, graphic design, and content presentation nudges our imaginations toward particular types of context (Student Computer Art Society, n.d.). D-Tech pings at us, and is increasingly being developed to draw on our emotions (Alasaarela 2017). The website blinks at us, and the news story expends 80% of its words on the 2% of the protesters who were violent.

The speaker who is crystal clear about the context being created, on the other hand, empowers us to query that context. In the second version of his story, Bob described the helicopter hangar's parts strewn on the ground. Bob's superior officer now has the opportunity to both understand the context Bob is suggesting and to question it.

We can also learn even more about the context by studying history or obtaining multiple sources of information. A friend with the opposite political perspective is invaluable. We may *dislike* what we hear, but we'll identify more methods for testing our understanding of the world than we currently imagine.

We can similarly improve our ability to manage context that our imagination subconsciously adds. If we are receiving statistical analysis of a study, for example, we can ask, "What image of the actual experiment do those numbers conjure in my mind?" and "Would other images be consistent with the equations I see?"

Our subconscious also adds bias based on whether or not someone is on "the other team." In one study, researchers transparently divided a group according to a random number generator—and what happened? Participants immediately begin treating those on "the other team" with less care (McRaney 2018).

By understanding the impact of such "tribal storytelling," we can more effectively evaluate content from and about people who are different from us. We naturally tend to be sexist, racist, homophobic, and anti-Semitic. We may even have an inherent bias against people who wear knitted sweaters.

However, we can remove the impact of our biases. We start by accepting that we subconsciously evaluate the world in the context of tribes. We can then treat content from people from "other tribes" with greater care, evaluating information from them more thoroughly, and thus ending up better informed (Nordell 2017).

In fact, in today's D-Tech world, understanding storytelling must be a standard for the development of our young. Such understanding goes beyond teaching basic logic to include our tendency to infer well beyond the facts we hear. To that end, "There has been a proliferation of efforts to inject training of critical-information skills into primary and secondary schools (Lazer, et al. 2018)." We should celebrate such initiatives while ensuring *we* take the actions necessary for those efforts to be successful.

TEACHING OUR MACHINES

Third, to support constructive storytelling we humans must take responsibility for the stories our machines tell—when designing, researching, and implementing new software.

In physical product engineering, a lead engineer vetoes specifications that would cause application problems. Such engineers creatively approach the impact of specifications on real world application. Imagine, for example, the specification:

"When the vehicle is traveling 20 miles per hour on a flat surface, the brakes must stop the vehicle within a two-mile stretch."

This might result in the implication:

A slow-moving vehicle is chasing little old ladies across mall parking lots. Mall cops are chasing the vehicle shouting, "Stop, Herbie! Stop!" (Figure 6.3).

Such a specification is eliminated before design begins—and then specifications that survive such preliminary evaluation are tested, again, upon completion.

Professor Cathy O'Neil highlights the importance of the equivalent quality control process for software engineering, based on a list of each stakeholder's concerns. For example, use of software to determine potential for prisoner recidivism in assigning parole decisions must be appropriate for the concerns of the public, the judge, and the convicted prisoner (O'Neill 2016, 23–27).

Once built, human users must understand their machines' approach—in order to proactively spot likely problems. For example, a colleague created software that recommends movie production budgets based on the script. The software does an excellent job—but does not consider whether a particular movie is a sequel. Rather than accept a $5 million budget for a poorly scripted *Star Wars* sequel, the human must adjust for the context that the software misses.

As discussed in Chapter 2, this does require that machine logic is transparent. Humans are the only ones able to understand context. Machines cannot. Thus, we humans must be able to oversee machine decision-making.

And, finally, we must recognize that such efforts focused on direct human control of machines will be incomplete. Our D-Tech tools, notably search engines such as Google, are used in massive numbers of situations. It is impossible, in every instance, to precisely program such tools to appropriately respond to the specific context of their use—or to make their logic appropriately transparent.

Figure 6.3: A car designed with weak brake specifications
would be a nightmare in a mall parking lot. We need to
avoid such mis-specification in all technical products—
even when the future problems are less obvious.

It is for that reason that we must focus, first, on the humans. *We must understand the process of storytelling, and remain vigilant about our machines.* The quality of the stories we tell, both directly and through our machines, must put our reputations at stake.

As a society, we face an inherent, fundamental tradeoff, however. The more history we have about an individual or organization, the more fairly we can award an appropriate reputation. Yet, the more information we have, the less privacy we respect. Today, the European Union and China are largely moving in opposite directions related to this tradeoff.

As part of the EU's General Data Protection Regulation (GDPR), the EU is ensuring individuals can request "erasure." Organizations so requested must, in many instances, erase all records associated with the requesting individual (Greenall 2018).

Such erasure is a powerful support for privacy. Yet, erasure is also a barrier to effective reputation development. Suppose the racist Satoshi spends much of his time posting racist epithets on the Internet—then requests erasure, puts on a suit, and asks, "May I please have a job, Sir? May I have a job, Ma'am?" Such options for erasure weaken the ability to reinforce constructive behavior.

China, on the other hand, is experimenting wholeheartedly with reputation. Chinese city governments are actively developing "social science credit laboratories" using detailed citizen monitoring (Algedi 2017). "Chinese security cameras use facial recognition to identify each person and then overlay . . . details such as age, gender, and clothing color (Smith 2017)." Such monitoring currently records actions as "petty as illegal parking (Algedi 2017)."

In the city of Rongcheng, China, you may be surprised by a banner flying outside the train station one morning—with an image of your face on it! You were just selected by the city government as one of the "best behaved" citizens (Mistreanu 2018).

Are we living in a dystopian movie? No. It's real life.

As a society, we have a tradeoff. Each option has ugly implications. The stronger we reinforce reputation, the more constructive our storytelling. Yet, the stronger the reputation, the greater the power of the organization deciding what *good reputation* is. "Who will guard the guards themselves?" ("Quis custodiet ipsos custodes?" 2018).

Every society must make its own decision—and we deserve to have the debate about what choice we want. In many ways, we should feel like we are back in 1787, the year the U.S. Constitution was written.

The focus on humans and storytelling also does not eliminate our ongoing focus on the validity of our facts. Today, D-Tech supports lying better than any technology in the past. We see artificial videos of the actress Gal Gadot in compromising situations (MacDonald 2017). Adobe offers a new software that "speaks" any sentence we suggest in the voice of any real or imaginary person we desire (Lozano 2016).

D-Tech, on the other hand, also supports fraud detection just as powerfully. D-Tech can help identify art forgeries by comparing brush strokes (Cascone 2017), and can effectively "prove" when and where a picture was taken (Omohundro 2018).

The historical battle between truth and lies will continue. When rationally analyzing our situation, we must have accurate content. Truth still matters.

Yet, storytelling matters far more importantly than in the past. Our understanding of the world around us will determine—far more than ever—our stories, and the reputational systems that support constructive storytelling.

Once we understand how our world is operating today, how do we coordinate our decisions as needed by a society of specialists?

HOW MUST WE STRENGTHEN COORDINATION? SOCIETY-SERVING LEADERSHIP

In most cases, we cannot, on our own as individuals, transform our silky Ferrari or utilitarian Honda into a *Ferronda*. We must collaborate with others; we'll install the Honda engine, but others need to change the transmission, shocks, and tires to match.

As we discussed throughout the book, we have leveraged D-Tech analysis to create an amazing diversity of options and specialists. The impacts of our decisions are deeply interrelated with decisions made by many other people and machines. We each oversee only small elements of the large, interconnected systems that deliver what we want and need. As a result, we must collaborate. Unfortunately, as evidenced throughout the past few chapters, too many of our collaborations have failed.

How do we solve the collaboration challenge? By learning from history.

Throughout history, how have great leaders aligned people's efforts to perform superhuman feats? Through leadership. At its core, the collaboration challenge we face today is much like the problems great leaders have solved in the past.

Yet, today's leadership challenges also differ from many across history—the scope of our collaborations is invariably broad, and powerful leadership most consistently derives from business. We need to embrace these shifts, learning the power of society-serving leadership. Such leadership frequently features key targets aligned across organizations. We need to develop confidence in such leadership.

We deliver society-serving leadership by expanding on our core leadership approaches.

THE CORE LEADERSHIP CHALLENGE

Leaders bring people together to make a difference. From the leaders of men in caves to the great Agricultural Societies to business today, the basic challenge has remained the same. And leaders have had available the same three basic tools: vision (setting of *purpose*), direction (*planning*), and delegation.

Omarr Bashir, CEO of Heritage Sports Radio Network (HSRN), describes how he's used the tools of leadership to build his business (personal communication, July 17, 2018):

> *Our purpose at HSRN is to develop bridges to the Historically Black College and University (HBCU) community. We build those bridges as the preeminent voice of HBCU sports.*
>
> *The concept of building bridges may sound odd for a sports broadcasting organization. I mean, 'Is sports really the best route to a broad community?' Yet, for the HBCU community, the answer is "yes."*
>
> *HBCU sports competitions are community celebrations. More than half a million people attend the seven highest-profile HBCU sporting events and more listen on the radio. It's not uncommon to have more people tailgating in the parking lot than attending the game. When you combine all of the HBCU competitions, we're talking audiences in the millions.*
>
> *Everyone is coming together for these games. You can imagine a big Thanksgiving celebration. We call them "family events"— but they're broader than that, including immediate family, broad relatives, friends, and sometimes even visitors from abroad with nowhere else to go.*

That's what we have with HBCU sports—the entire African American community coming together for a celebration. We HBCU alumni consider ourselves graduates of all 101 HBCU schools, not just our own universities. As a result, we are all likely to attend a nearby game. And, with that strong nucleus of neighborhood leaders attending, our crowds draw others from the local African American neighborhoods as well. We say everyone in the African American community is one-degree of separation from an HBCU alum—but that probably underestimates the connections. Most people in the African American community are one-degree of separation from multiple HBCU alums.

At HBCU sporting events, we're all tailgating, reconnecting with old friends, and making new ones. We talk about what's important to the community, and community perspectives get solidified at such events. What's our view on Starbucks shutting down all their stores for racial sensitivity training? You know that was a big topic at recent games. So was the importance of academic scholarships.

The crowds participating in these events develop a shared understanding. We take these perspectives and understandings back to our neighborhoods and we spread them further. Everyone participating in these events feels a kindred spirit, and that's powerful.

When we founded HSRN a decade ago, we couldn't develop all bridges at the same time. Our initial plans focused on building basic business infrastructure while strengthening bridges within the HBCU community. We built the physical infrastructure to broadcast nationally, developed sports announcing standards, and expanded a network of broadcasting ties. We supported this growth with traditional advertising.

Even when just starting out, however, our Purpose remained front and center. To build bridges, we needed more than the respectful—but distant—greetings at sporting events that the billion dollar broadcasting corporation receives. At HSRN, we greet the HBCU community with hugs and high fives—we're all about connecting and celebrating.

We also go beyond sports, initiating every broadcast with Heritage Salutes, an interview with a notable HBCU alum. The pride in our HBCU alumni is high—from Oprah, to Martin Luther King Jr, to Toni Morrison, to P Diddy. Recognizing our ties with such people strengthens our community bonds.

We've done well through such efforts, rapidly developing a national presence for HSRN. Yet, we've also benefited the HBCU community, and our employees are proud of that.

Everyone at HSRN loves sports—but we also understand that we play a key role in society. At HSRN, we feel pride in our purpose of building bridges between people. African American suicide rates are less than half the rate for Americans overall. Even a single suicide is too many, yet African Americans' low suicide rate is still something to celebrate. We like to think that HSRN's work plays at least some role.

Many of our advertisers feel pride in being associated with such success, as well.

Such leadership brings people together to accomplish more than they ever could as individuals. The core of such efforts is Purpose-Driven Leadership, which helps individuals coordinate their actions around a high-value purpose: combining vision (an organization's *purpose*), direction (clear *plans*), and delegation to engage and align participants.

Figure 6.4 (from left to right) depicts the process through which Omarr Bashir developed his leadership approach.

On the far left is Purpose. An organization's *purpose* is a bold affirmation of its reason for existence—a choice of how to serve society (Dvorak and Ott 2015). Heritage Sports Radio Network (HSRN), for example, could have determined that its purpose was to entertain. It could have chosen to be a variant of the sports network, ESPN, which focuses on "meeting sports fans wherever they are" (Freedland 2017). Instead, HSRN's purpose is to develop bridges—a purpose with even greater value to society.

Purpose-Driven Leadership

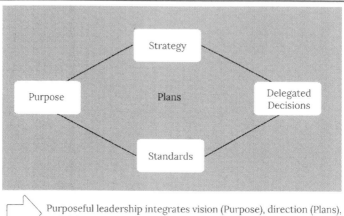

Purposeful leadership integrates vision (Purpose), direction (Plans), and delegation to strengthen coordination.

Figure 6.4: Purpose-driven leadership integrates vision, direction, and delegation.

From the purpose is derived the middle set of boxes: the *plan*. A well-led organization develops *plans* consistent with its *purpose*. Plans consist of a combination of:

- strategy: concrete actions, and / or
- standards: a consistent approach to handling particular types of situations.

The best plan is to deliver the most attractive immediate outcome, while being consistent with long-term delivery of the *purpose*. Especially early in an organization's existence, plans cannot deliver the entire *purpose*. HSRN couldn't initially develop all of the bridges to the HBCU community. Such early plans are a stepping-stone to strengthening the organization and developing the capabilities that will later deliver the purpose.

Execution of such plans is delegated to employees, which is on the far right side of Figure 6.4. Employees execute in the reverse order, moving from right to left in the figure. Employees apply three filters when considering initiatives.

The first filter is profitability. Business employees understand that they are conducting business, and focus on high-profit-potential initiatives.

Initiatives that pass the first filter face the second filter: the plan. Actions consistent with the plan, for example serving clients in target regions, receive highest priority.

Finally, execution of initiatives proceeds according to the third filter: consistency with the *purpose*. HSRN ensured it maintained strong boots on the ground so that connections with the HBCU side of the bridges remained strong. When HSRN has two finalists for a new position, they hire the one better able to help deliver on the Purpose over the long term. Perhaps that means the employee that better knows the HBCU community, or who brings a unique skill set for building bridges.

Over time, the ability to fully deliver on the group's purpose expands. Hundreds and then thousands of decisions are made— all consistently according to these filters. The organization's understanding of the hurdles to delivering on their purpose grows; the organization invests in knowledge and capabilities necessary to overcome such hurdles.

Every organization, including every business, should be focused on purpose-driven leadership. Larry Fink, the CEO of Blackrock (which manages more than $6 trillion in investment funds [Sorkin 2018]), says "society is demanding that companies, both public and private, serve a purpose . . . Without a purpose, no company can achieve its full potential (Fink 2018)."

Larry Fink is correct. Certainly, society benefits when groups of people coordinate on delivering something of value to society.

Yet, investors benefit as well. In fact, all investors should *demand* their businesses apply purpose-based leadership. Where's the investor value?

- **Employee Engagement:** We desire to "make meaning rather than make money (Kawasaki 2014)." Employees at truly purpose-driven organizations are more engaged, innovative, and effective (Knippenberg 2001).
- **Customer Attraction:** Consumers want to do business with those serving society. We're not just buying products and services any more—we want our brands to benefit the world around us.
- **Employee Alignment and Value Creation:** Leadership builds collaboration—avoiding the failures described in the previous chapters. And, when such collaboration solves one of our many societal failures, value is created. Think about the *value* created if the American healthcare industry efficiently helped us live extended, productive lives! Consider the value of a food industry that delivered nourishing food. Every investor should be salivating!

Some businesspeople and investors squirm when they consider society-serving leadership. They are not used to recognizing that a long-term focus on delivering a society-serving *purpose* is also in their personal interest. Those businesspeople don't recognize that such a focus on society-serving purpose complements their focus on returns—because returns provide the *first* filter for their actions. Executing according to purpose only occurs if that first filter is successfully delivered.

Let those businesspeople squirm. Take their share of the value created.

TODAY'S LEADERSHIP CHALLENGE

Today's D-Tech leadership challenge differs from those hurdles we have overcome throughout our history. Discussing the purpose of a single organization is not enough. Active support from our Government or Cultural leaders can help, but is frequently insufficient to solve our deep-seated social problems. Business leads today's society. Thus, <u>business-driven</u> purpose must be aligned across *all* <u>organizations</u> needed to fix our broken Ferrondas.

The next stage in HSRN's development, as described by Omarr Bashir (personal communication, July 9, 2018), illustrates the leadership challenges of today:

> *We need to build broad bridges between the African American community and corporate America—and that will require alignment across many organizations.*
>
> *Forbes magazine says the African American community controls $1.2 trillion in spending power. That's more spending power than the 15th largest economy in the world: Mexico. Yet, the African American community isn't just a neighbor to the U.S.—it's part of the U.S.*
>
> *Nonetheless, miscommunication seems almost constant between the African American community and corporate America. Human Resource professionals complain about challenges recruiting the best African American employees— and then keeping those with highest potential. Advertising executives are too frequently surprised by African American community responses to their corporate messaging. And marketing specialists leave too many selling opportunities undelivered because they're unsure how to execute.*

> *We need a broad, stable bridge between these two communities. The HBCU community is a natural foundation for such a bridge. As I said earlier, the HBCU community is inherently an entry point to the African American community.*
>
> *In addition to sponsorship, we see breakfast conversations, discussion events, and more—all to facilitate quality, two-way communication between corporate America and the African American community. Each corporate participant in such events is brought in as a welcome guest of the highly respected HBCU athletics community. Such an experience is the equivalent of a homestay in the African American community for those organizations serious about engaging.*
>
> *Meanwhile, developing that bridge through the HBCUs supports those institutions' delivery of the African American community's highest priority—education.*
>
> *Many organizations must collaborate for such a bridge to be strong, however. Our current efforts include aligning dozens of HBCU institutions, members of corporate America, and others. So many people have approached us with goodwill towards this venture—yet all of those positive sentiments can only have the intended impact when they share an aligned purpose and the outlines of a high-quality plan.*

Leadership brings people together to align their efforts toward a common purpose. In the case of modern society, more people need to be aligned than ever before—and that alignment needs to facilitate more complex collaboration than ever before. So, how do we execute?

- We ensure the purpose is clear and plays an active role in decision-making.
- We secure a plan that provides clarity to a larger segment of society—while allowing that mass of individuals the flexibility to react according to standards that ensure a consistent approach.

- In the bulk of cases, we connect our efforts with Business.

CLARIFYING AND ENERGIZING PURPOSE

First, *purpose* must be clear and play an active role in decision-making.

Wherever large numbers of people must collaborate, that group must be clear about the *purpose* of their collaboration. It is only through such clarity that they will be able to coordinate their decisions. For American healthcare, for example, we need clear, aligned *purpose* shared across healthcare-related organizations. In particular, we need a clear *purpose* that plays an active role in decision-making at the federal level: aligned across the states, that will be implemented across Government, Business, and Cultural institutions.

Rather than build alignment across organizations and geographies around meaningful purpose, however, we debate an almost meaningless question: what's the best healthcare system in the world? Pundits point to Canada. Or Germany. Japan. Singapore. Israel!

Such a question is akin to asking, "What's the best car in the world?" If I'm going to the supermarket to buy a watermelon, I don't want the Ferrari: I'd have to sit on the watermelon on my way home because there's no trunk. On the other hand, if I want to preen myself and go out on a Saturday night, I don't want the Honda. Different people value different aspects of their cars—which is why different people buy different cars.

Yet, no matter what type of car is being built, the engineering team is clear—up front—about their shared purpose. For the team designing a Honda, the purpose is to create a utilitarian, low-maintenance vehicle for (dull) family transport. For the

Ferrari team, the purpose is to create a sexy, sleek, sporty vehicle to be drooled over by non-owners.

In contrast, our healthcare team is building an ever-more-complex Ferronda. We try and optimize individual elements of that healthcare system, rather than delivering clear missions, each aligned with the overarching purpose.

Our healthcare workers *want* their work to be purposeful. Seventy percent of respondents in a recent Harvard Medical School survey said they hoped their future work would be more mission driven (Silver and Bhatnagar 2017). I'm confident our bureaucrats and regulators *want* the rules and regulations they create to improve our healthcare system. Yet, to make things better, and make a difference, we need alignment—we must have shared *purpose*.

Today, such clarity of purpose is missing.

Once we have such clarity of *purpose*, we need to consistently use that statement of purpose in decision-making. Can you find your organization's official *purpose* statement on the website? Probably.

Can you find the implications of that *purpose* statement's words in your business's, government's, or cultural institution's decision-making? Less likely.

Unfortunately, window dressing doesn't change the world. Action does. And we have too little aligned action. Ninety percent of healthcare workers in the same Harvard Medical school survey referenced above suggest administrators are definitely or sometimes less mission driven than their employees (Silver and Bhatnagar 2017). Presumably, those administrators are unaware of, or too frequently ignoring, their organization's purpose (as well as simply missing the non-existent federal purpose statements).

ENGAGING PLAN

Second, aligning people's efforts toward a common purpose requires that plans be engaging. While too much of our work lacks a clear, high-level purpose and plan, many existing plans have far too much detail. Plans must respect the flexibility individuals need to successfully respond to their actual situations.

With D-Tech, our world is becoming increasingly complex. Yet, too frequently, we get caught up in the highly specialized data made available through D-Tech without understanding the complex, interconnected contexts that lead to failures. As a result, we create new standards and regulations that respond, piece-meal, to specific crises and failures as they emerge—and then we store those standards and regulations in our digital vaults that never forget.

We also should study how amazing successes were achieved, as described in Braithwaite, Wears, and Hollnagel (2016). Engaged employees deliver spectacular success when their *purpose* is clear and they are provided the flexibility to react constructively to the byzantine real-world situations they encounter. Failure becomes increasingly likely when disengaged employees are tied down by excessive planning.

Our healthcare workers, to illustrate, are inundated with plans featuring *too many* standards: standards for checklists, approaches for filling in the non-checklist aspects of checklists, and for checklists focused on appropriate storage of other checklists.

Why are more than two in five physicians (42%) burned out while 15% are clinically depressed (Ault 2018)? Because the plans those professionals must execute are far too detailed. Those professionals lack the flexibility to work effectively.

Across the United States, 70% of American workers are disengaged in their jobs (Maxwell, Locke, and Ritter 2016). Few of these men and women are begging their managers, "Can you please create just a few more rules and standards for me?" Most are watching for the seconds passing to bring them to the end of the day—where they can go home to avoid all of the detailed plans and standards.

BUSINESS DRIVEN

And, third, Business must generally lead collaboration toward a common purpose.

What is happening across our Government and Culture? As discussed in Chapter 4, it is increasingly challenging for Government to think clearly. Government must now gather global information to regulate our nation, understand many specialty languages, and operate as a democracy despite increasing potential for voter manipulation. In other words, Government must operate in a maelstrom of noise and confusion, too frequently dependent on translation by special interests to understand our world. In most instances, Government is thus poorly positioned to lead well.

As discussed in Chapter 3, our Culture is struggling to help us align contradictions and provide direction through our haze of complexity. Thus, our Culture, also, is poorly positioned to lead.

Culture will play an important role supporting not-for-profits as the wealth entering not-for-profit foundations explodes (Fleishman 2009). Yet, not-for-profits have at least three inherent weaknesses: their solutions tend to be incomplete, they are too frequently associated with misunderstanding the nature of society's problems, and they lack Business's resources. Need a shiny building with a nameplate, for example? Not-for-profits are likely to find funding. Need that shiny building maintained? Not-for-profits are less likely to find the cash.

Not-for-profits will continue to provide amazing good for society. However, for much of the deepest change in society, not-for-profits will need to align across organizations with Businesses.

Business has the true power in 21st-century American society. Business winners have more resources and greater discretion about how to apply those resources than ever before. Moreover, as we've discussed in this Chapter, Business's self-interest lies in providing Society-Serving Leadership.

Other examples of Society-Serving Leadership exist today than what we've discussed thus far. The cocoa industry, for example, exhibits Society-Serving Leadership. The world's largest cocoa and chocolate producers (including competitors Barry Callebaut, Cargill, Hershey, Mars, and Nestlé) provide Society-Serving Leadership for cocoa farming. These businesses share aligned *purpose* with several non-governmental organizations and others around improving the sustainability of cocoa farming.

The coordinating cocoa Businesses continue to compete in most respects. Nonetheless, the group's collaboration has successfully improved the quality of life for many Ivorian cocoa farmers as well as improving the quality and quantity of cocoa available for chocolate. As Taco Terheijden, Cargill's Director of Cocoa Sustainability, highlights, "By sharing these goals together with our partners, we can align to accelerate transformation towards a more sustainable cocoa supply chain and better businesses for all involved (Cargill 2017, 6)."

Yet, as evidenced throughout this book, we benefit from too few such efforts. Our leaders must earn the label "Society-Serving Leadership" for what they do. They must integrate a systemic view of service to society with purpose-driven leadership. That purpose must frequently span across organizations, it must support engaging plans, and, in many instances, be business-led.

CONCLUSION: SOLUTIONS TO OUR ILLS

Five. Four. Three. Two. One. Rumbling fills the air. The rocket lifts from the launch platform. Awe fills the crowd.

Time and again, NASA's Jet Propulsion Lab (JPL) delivers spectacular missions despite seemingly overwhelming challenges. Missions have successfully traveled to the moon, Saturn, Mars, and asteroids, while others constantly orbit the earth. JPL scientists fill our night skies with precision-executed space missions.

Armies of people are involved in a mission: physicists, chemists, artificial intelligence specialists, mechanical engineers, and more. Each of these rocket scientists must achieve utmost precision in their individual tasks—and those tasks must coordinate perfectly with one another. Misjudge liftoff by a millimeter and the rocket may miss its target by millions of miles; there may not be fuel available to correct such a mistake. Technology must be stretched to its utmost, without room for failure.

After liftoff, scientists fill the control room, peering over each other's shoulders for a glimpse of the most recent measurements, images, and recorded sounds. Engineers must cram every crevice in the rocket with delicate, cutting-edge technology to fulfill as many of these scientists' wishes as possible. An ounce of weight wasted on one piece of equipment is an ounce unavailable for another—and an ounce unavailable for more fuel to correct any errors.

The scientists at JPL are essentially tossing a tin can millions of miles to hover inches outside some outpost from which they will receive groundbreaking scientific information. Yet, such virtually impossible missions are regularly successful. How?

JPL scientists are *expert coordinators*. With a clear common purpose, they constantly reinvent their systems to better adapt to challenges. Their work is organized in three key aspects: a powerful structure for information, focused storytelling, and effective leadership that brings the diverse group of specialists together.

JPL scientists effectively *structure their information*. Take this scenario, for example. While Electrical Engineer Brad fine-tunes the electrical connections between the rocket's battery and every single piece of equipment on board, Systems Engineer Tabitha is balancing weight across the entire rocket. Tabitha has learned that the radio will need to be 500 grams heavier than initially expected due to a miscalculation. She must decide whether to reduce the amount of scientific equipment loaded onto the rocket or to spend more on a bigger engine capable of carrying the extra weight.

To make her decision, Tabitha needs the weight of each area of the rocket. JPL helps Tabitha make that decision by structuring its information with a semantically rigorous systems engineering approach (Jenkins and Rouquette 2012). These semantics clarify that the weight of the wires associated with an area such as the solar panels are to be included in that area's total weight. Tabitha can therefore receive the general information on weight by area (relevant to her question) without sifting through the detail of Brad's work.

JPL scientists are *exceptional storytellers*. Several times during a major project, NASA engineers will come together for project reviews. They know their mission depends on every team's flawless work. Even a tiny mistake could send their space probe millions of miles off course. What has taken months to design must be clearly communicated in minutes. The story must be vivid enough to convey the project's exact challenges as well as the engineer's plans to solve them. Brad and Tabitha must clearly understand these facts and be provided enough context to rigorously challenge the communicated approach.

Every organization has reviews. What stands out about JPL reviews is their intensity. The team must dig as deeply into each other's work as possible. Through stellar storytelling, they are able to imagine the risks identified by team members—and those not identified.

Thanks to *excellent leadership*, the work of all JPL specialists effectively comes together to fulfill one coherent mission. Leadership is clear: the goal of the mission, the type of space probe that will be built, and the core equipment that will be necessary to create that probe. The JPL team doesn't build a *Ferronda*. They choose ahead of time to focus on one specific vehicle and they ensure the system elements all work together seamlessly. In other words, JPL leadership communicates a clear common purpose; provides simple, engaging plans for achieving it; and promotes collaboration toward this common goal.

There is a powerful lesson in JPL's approach. Massive technology, complexity, risk, and change—it can all be mastered. Collaborating in such an environment means applying rigorous tools that allow cutting-edge specialists to work side-by-side with generalists. Information must be structured, the art of storytelling must be mastered, and specialists must be brought together to collaborate towards a common, higher goal with effective leadership.

Outside of JPL, however, we frequently ignore the true, systemic nature of the challenges created by complexity. We look only at our own piece of the puzzle, we dig deeply into the relevant information linked with our own piece, and we forget to effectively bring together the insight of all of our specialists.

Digital Technology offers spectacular potential. We have the technology with which we could solve most of society's problems. D-Tech and our existing Culture, Government, and Business/ Economy nourish the specialists we have been developing for centuries. With improving technology, these specialists are developing an ever deeper and more detailed view of our world.

Yet, we are *Drowning in Potential*. New challenges are emerging and we have not strengthened our ability to bring people and knowledge together. More than ever, we must collaborate to make an impact. Yet, instead of collaborating, we are miscommunicating and misaligning. Our Culture is causing us despair rather than connecting us. Our Government is setting us up for failure rather than protecting us. And, our Business and Economy are not delivering many of our highest-value needs. We must change—or increasing segments of our society will fail.

We *can* fix society. As each technology revolution has progressed, we've witnessed some societies reinvent the tools for collaboration and apply them broadly. We've learned how to support markets through the creation of appropriate standards and supporting rules. Mathematicians at JPL and elsewhere understand how to create the clear and precise methods that could underpin *Markets for Precise Concepts*.

We can instill an environment in which the best storytellers, and the best story listeners, are rewarded for their efforts. There is, arguably, nothing more human than telling a good story: compelling, clear, and insightful. We can develop strong reputations where we need them, while preserving privacy and confidentiality where we don't. We can develop a reputation system that will help teens, for example, recognize that they will be rewarded with good reputations when they learn about different cultures. And, those teens won't have their futures destroyed just because they make a single mistake. In short, we can enhance understanding through *Reputation-Based Storytelling*.

Around us, wonderful examples of clear, powerful leadership set an example for the future. Leaders at JPL don't build Ferrondas; they build clearly specified, effectively operating spaceships. Business leaders are already responsible for the most powerful institutions in America. Like the leaders in the world's cocoa and chocolate industry, our business leaders can extend their influence to solve society's most challenging problems. There is profit to be made there. When society has a problem, value is being delivered to no one. If our powerful business leaders lead the charge toward problem solving, we can certainly share the value of the fixes with them.

Using JPL's structure as a model and an inspiration, we can redesign society. We have the technology to develop amazing space missions. This same technology can cover improvements across every aspect of our society. But such spectacular successes arise only when we focus on a common purpose and we collaborate to create appropriate solutions. *Technology* doesn't solve problems. *People* solve problems.

My final memories of my father include many tubes, cords, and beeps. But "Please don't let me die alone," he had asked us. To honor his request, my family and I turned off the technology. We connected with each other, we said meaningful prayers together, and we said goodbye.

As leaders, as parents, and as members of society, we have a choice. We can give up. Or, we can change our approach: stay vigilant, identify the major failures that surround us, and focus on collaborating with others to fix them. American society *can* cut through technology's unintended impacts. We *can* survive Digital Technology.

I still feel worried about American society. I don't know if our society will flourish or fail. I do, however, have hope. Please join me in reinventing collaboration across America. Let's create the society we want to leave to our children—and let's benefit ourselves and others while doing so.

Please join me in reinventing collaboration across America. Join my list of subscribers at RodWallacePhD.com, or follow your own path. Let's create the society we want to leave to our children—and let's benefit ourselves and others while doing so.

THANK YOU!

I hope you enjoyed Drowning in Potential.

*Please <u>write a short review on Amazon</u>
to help others find this book.*

ROD WALLACE

RODWALLACEPHD.COM

ACKNOWLEDGMENTS

The fingerprints of dozens of people reside in this book. From concept to copy to cover, I've received a wealth of assistance.

Uncountable ideas were born from concepts shared by **STEVE OMOHUNDRO**. Steve has been an insightful mentor as well as my loving brother. I am forever thankful for my years of learning from him. Extensive discussions with **LLOYD NIRENBERG** and **RUTH FISHER** further helped flesh out many of these ideas. Lloyd's gritty comments consistently provided the unvarnished feedback I needed to target improvement. For Ruth, a special thanks for her energy and enthusiasm for this topic, and her contributions to the book from its early stages.

A special thanks also to **BOB MERKLE** for the extended sessions discussing digital technology's unintended impacts on business today—and for helping identify the origins of these impacts in the late 1960s and 70s. **JILL NEPHEW** played a similarly critical role in developing a simple yet precise way to understand how the information we receive influences our thinking.

VIBHAT NAIR and **JING WANG** have been the deepest of friends. Their contributions only begin with their unique, perspectives on, and wonderful challenges to, the ideas in this text. Their stories born of work in America's largest corporations add a critical depth to the stories told here. Even more than their stories and ideas, their friendship has been exceptional—a critical source of warmth during the coldest, darkest days of writing.

ACKNOWLEDGMENTS

The prose is as much the work of my editors as me. **RACHEL MCCRACKEN** knew nothing about me when she began editing. Yet, she threw herself, her husband (Thanks Adam!), her friends, and her professional contacts into this book without reserve. **QAT WANDERS** did the same. From helping me be more concise to late-night calls to suggest others who could help—Qat was amazing.

ANN MILLER provided more than just editing insights. Structure, grammar, and clarity—Ann's teaching have helped me communicate more clearly in all dimensions. Over the course of months, Ann freely shared her time, insight, and experience. She's made innumerable suggestions about how I could write more clearly, find my audience more effectively, and find others passionate about my topic. A special thanks to her.

For **MICHAEL O'FLAHERTY** also, my thanks is for more than editing. I feel this book is his as well as mine. Researching, indexing, editing. His efforts have been monumental. How many times did I see an email from Michael arrive at two in the morning, I responded at that time—and then found yet another response from him waiting when I woke up? A thank you for Michael.

A thank you for those others who also provided such critical comments about the text and concepts along the way. **BRIAN POI, RICK GROSS, HEATHER** and **SCOTT GILBERT, JOS DELOOR, LACEY COLLIGAN, RANDY KAMEN, IRA GOODMAN, PATRICK MCGUIRE,** Doug Stevenson, **BRAD CLEMENT, ROEL KLEIN, JESSICA HOLDEN-SHERWOOD, DR. HOAG, PAT MCGUIRE, JIM SHULMAN, JIM SHULMAN, JIM LOLA, STEVEN SWERNOFSKY,** and so many more. Thank you for your time and thoughts—especially when the early drafts you read still required so much improvement.

What we see in these pages is the result of more than just words. **IDA FIA SVENINGSSEN** at IF Design

(IdaFiaSveningsson.se), and **KAREN FERREIRA** and her team at GetYourBookIllustrations.com were amazing partners. They developed the style of the illustrations and the creativity that allowed so many of the scenes in this book to come alive. **MICHELLE VILLALOBOS'** support in clarifying direction (SuperStarActivator.com), **CIARA PRESSLER'S** brand development (CiaraPressler.com), **RYAN HAAGEN'S** videography, **PENNY SANSEVIERI** and her team at Author Marketing Experts on marketing support (AMarketingExpert.com), Jenny Bartoy at BrightEdits.com for amazing copy editing, and **LISE CARTWRIGHT'S** perspectives on the publishing industry. For your yeoman's work—Thank you!

Many have helped me learn to collaborate effectively with this bevy of professionals. My colleagues in the 2018 National Speakers Association (Philadelphia Chapter) Speakers' Academy bear special note.

The incredible tutelage of **GERRY LANTZ** has been exceptional—and he has yet to tire of my litany of questions. Thank you!

For the ongoing support, perspective, and insights from **RITA WILKINS** and my other colleagues—thank you, as well! And, to **RANDY KAMEN** and my new-found colleagues in "Write, Self-Publish and Promote Your Book," my learning from you is just beginning, yet already exceptional. Thank you.

More than thank you, perhaps bless you is the appropriate phrase for **TOMMIE ALMOND**. Motivational support, countless hours editing, and extra help taking care of my eldest son just begins to describe what she has done for me. You chose to become a pillar upon which I stand. For you, I send thanks.

And, thank you to my family. **MY WONDERFUL WIFE, MING**, is the only one who's seen multiple drafts of every section of this book—and all two thousand pages that sit on the editing room floor. Thank you for that, but even more, thank you for your love, support, and care—for me and the boys. **AND TO MY CHILDREN DEVIN, RAYDEN, AND REUBEN**—you are the source of my desire to make a difference. You are the source of love . . . and you've also been the source of the bits of chocolate I've enjoyed throughout the research and writing process.

THANK YOU ALL!

ENDNOTES

CHAPTER 1

1   A society is clearly more successful if it delivers more of the needs and wants of every citizen who participates in it. We do not judge how the impact of the distribution of wealth among citizens impacts the measure of success.

2   In fact, Richard Kelly says there is only one exception to the parallel between the evolution of technology and living organisms: technology never becomes extinct. We can always bring back technology that is not currently being used. Of course, if geneticists are successful at bringing back extinct animals (see Zimmer 2013), perhaps even that difference is only temporary.

3   Kurzweil describes Moore's law in terms of exponential growth in information-processing technology. Exponential growth rate equivalent to doubling every two years is 40% per year.

4   Farmer and Lafond (2016, 648) find that "while . . . structural breaks happen, they are not so large and so common as to over-ride our ability to forecast."

5   To be precise, the authors cannot reject their null hypothesis that the technologies improve in a random walk with consistent mean and standard deviation.

6   For mathematical intuition, you can consider a process growing 10% per year from 100. The first year, growth is 10 units. The second year, growth is 11 units. The third year, growth is 12.1 units—each year accelerating.

7   This statement about number of parts is for illustrative purposes. More precise is that the number of dimensions on which we can improve a technology is constantly expanding. For cars, a potentially developing innovation is the shift to electric cars, for example, one benefit of which would be a reduction in the number of parts (Douris 2017).

8   There certainly have been eras in which technological development has slowed, e.g., during the dark ages. However, this slowdown was the result of problems in society rather than a result of less potential technological development. This impact back from society onto our development of technology is not a primary focus of this book, but certainly exists.

9   Peter Drucker (1970) focused on humankind's ability to manage the irrigation of the fields, while Harari (2015) focuses more broadly on agricultural technology. The point made by both authors is similar: human civilization was revolutionized during the Neolithic era which began as early as 15,200 BC and ended between

4500 and 250 BC, when humankind began to develop and use metal tools. In different parts of the world, this period, marked for its agricultural technology progress, began at different times and progressed at different paces ("Neolithic" 2018).

10 Frequently, these Governments were intertwined with religious institutions.

11 There are records that suggest at least preliminary concepts of uniqueness of the individual from the tribe existed even prior to the Agricultural Revolution. However, these are exceptions, and are unlikely to have incorporated the concept of "citizen," which was only needed once people outside the tribe became involved in commerce.

CHAPTER 2

1 The impact of snowy weather on auto accidents is an intuitive example of emergent impacts. Because of snow, car accidents are more likely. However, no snowflake pushes one vehicle into another. The snow causes the road to become slick and blinds drivers. For many accidents, you could explain, "The accident occurred because the snow made driving more hazardous." The accident was an emergent impact of the snow. You could also trace the origin of that same accident to, "The driver of Car A pressed the brake pedal too hard and ended up careening into Car B." The origin of the accident is not uniquely traceable.

2 For example, the physicians may want to prioritize the patient's participation in athletics or focus on a regimen that simplifies what the patient must do. Either such approach is likely to be most effective only with alignment among all involved.

3 The year 1970 is when, with the introduction of the first semiconductor chip was introduced, that challenged magnetic core chips (Computer History Museum 2015). Others have suggested a start date of 1965 or 1972. Using either of these alternate dates would have little impact on our analysis.

4 Data from "Hotels in Budapest" (n.d.).

5 Author: K Limkin, licensed under Creative Commons license.

6 Increased computer-processing power and new D-Tech programming approaches are responsible for these new capabilities. Many of the visual tasks of Digital Technology are executed using neural nets. See, for example, Goodfellow, Bengio, and Courville (2016), or the more recently discussed variant "capsules" (Simonite 2017).

7 Blockchain technology provides Digital Technology the ability to guarantee information provenance, for example (see Rode [2017a, 2017b]). Similarly, use of mathematical proof allows Digital Technology to guarantee that software meets specifications, i.e., that it doesn't make any flaws in its logic about the world.

8 For example, before Arthur and Lucy were in love, the two took a romantic boat ride. The Digital Technology interpreted them as being, "In boat," at that point. That statement doesn't even make sense as a reference to their relationship.

9 Author: Zizou Man, licensed under Creative Commons.

10 The game of Go is played on a board with 20 squares on each side, which means there are 400 squares in all. At the end of the game, each square belongs to either a black or white player. The first square could belong to each player, the second could belong to each, etc. Thus, there are 2 possible outcomes for the first square; 4 possible outcomes if the first 2 squares are taken together, with the possible outcomes doubling each square. 2, 4, 8, 16, 32, 64, 128 possible outcomes, . . . that number gets huge fast. By the 400th square, there are more possible outcomes than atoms in the universe.

11 A technologist may suggest an appropriate evolutionary algorithm to identify such 'offbeat' options, but making such an option work in the real world is another challenge.

12 A pixel is "one minute area of illumination on a display screen, one of many from which an image is composed" (Oxford University Press, n.d.).

13 Photo by Michael Schilling, licensed for use under Creative Commons Attribution.

14 Photo by Gordon Correll, licensed for use under Creative Commons Attribution.

15 See, for example: Yosinski et al. 2015; Kim, Malioutov, and Varshney 2016; Kurzweil 2016; Kim, Khanna, and Koyejo 2016.

16 We use 2013 as the start of the Artificial Intelligence era of the Digital Technology Revolution. That is the year Google's Deep Mind provided a single software that could learn to play multiple, different video games simply by watching them (Mnih et al. 2013).

17 This concept of Artificial Intelligence comes from Dr. Stephen Omohundro's (2008) definition: "*Artificial Intelligence* is the science of building machines, limited by scarce resources, able to take actions that achieve desired outcomes in a particular environment." Many other definitions of Artificial Intelligence exist; for 23 examples, see Marsden (2017). Many definitions other than those we've used are perfectly valid for their own purposes; each provides a unique perspective through which to view current Digital Technology progress.

18 Being provided all of the nuanced details on a subject doesn't help our analysis; it overwhelms and stresses us (Tartakovsky 2016).

19 We generalize the component parts to free up brain space. For example, if we're wrestling with how to organize a presentation, we may think of "customer 1", "customer 2", and "customer 3". We know full well that those 3 customers are materially different. However, when we're focused on how the higher-level task of structuring the overall presentation, we are liable to think of them as generic (Stockton 2014).

20 Description of the following information was provided by a Tinder spokesperson.

21 In addition, your tribe wins because you clear out the food another tribe would require to successfully invade. I.e., the strategy is evolutionarily stable (Liberman 2005).

22 This section based materially on Crawford (2017)

23 Glena M. Crooks, CEO Strategic Health Policy International, Inc., ex-Deputy Assistant Secretary for Health. Conversation by phone, Tuesday, March 13, 2018. For this count, a patient is one type of person, so is a patient's family member, pharmacist, lunch server, and hospital check-in staff.

24 Examples of types of people included in the count: plumbers, electricians, cable operators, gardeners, and painters.

25 The specifics of this example are masked to protect intellectual property.

26 Slowdowns occurred in Indonesia, Japan, Malaysia, North America, Pakistan, the Philippines, South Africa, and Vietnam.

27 Based on text provided by Ruth Fisher.

CHAPTER 3

1 Translated using transl8it.com:
"Are they really? :("
"You've almost been away for six whole weeks!! THAT'S INSANE." "Time flies yo."

2 Conversation with Suzanne Kaplan, March 19, 2018.

3 "In Chile, touching food with your hands is considered ill-mannered. Yep, even fries! In Brazil, too, pizza and burgers are normally eaten with a fork and knife (Boscamp 2017)." Also, "Norwegians eat burgers with (gasp!) a knife and fork. Norway's famous open-faced sandwiches are tough to eat without silverware, so they probably just decided to apply the habit to American sandwiches as well (Strutner 2017)."

4 "In the US, a firm, short handshake indicates self-confidence and (heterosexual) masculinity. A limp handshake by a man can be interpreted (usually wrongly) as a sign of homosexuality or wimpiness. But in most parts of Africa, a limp handshake is the correct way to do it. Furthermore, it is common in Africa for the handshake to last several minutes, while in the US a handshake that is even a few seconds too long is interpreted as familiarity, warmth and possibly sexual attraction (Borgatti, n.d.)."

5 See, for example, Hudson (2013), Simon (2017), and Trout (2017), respectively. Note the last reference is a discussion of the use of robots for sexual relations.

6 From Wikipedia: "Yellow journalism, or the yellow press, is a US term for a type of journalism that presents little or no legitimate well-researched news and instead uses eye-catching headlines to sell more newspapers. Techniques may include exaggerations of news events, scandal-mongering or sensationalism ("Yellow Journalism" 2018)."

7 Some exceptions were retained for U.S. Government propaganda, particularly around national defense.

8 The General Accounting Office (GAO), for example, delivered reports citing such activities as propaganda, and hence illegal. However, such reports were ignored, and, in at least one case, an official memo went out that the GAO should be ignored on this topic (Barstow and Stein 2005)

9 At face value, the Communications Decency Act of 1996 was passed "to regulate pornographic material on the Internet ("Communications Decency Act" 2017)." The original intent of the law is not relevant in terms of its impact on our Culture today.

10 Turnout figures are from the Statistical Abstract of the United States (various years), based in turn on surveys by the Census Bureau. On local turnout, see Verba, Schlozman, and Brady (1995, 69).

11 This data is reasonably consistent with, "According to the General Social Survey, between 1974 and 1998 the frequency with which Americans 'spend a social evening with someone who lives in your neighborhood' fell by about one-third (Putnam 2000, 271)."

12 We are spending increasing time with our families, including helicopter-parenting our children. However, it's unclear such relationships are any more healthy, on average, for society.

CHAPTER 4

1 A restriction is either an obligatory act needed to do something or get something (e.g., you must file form WX-175) or a prohibition from doing something directly (e.g., you may not wire money to country ____).

2 Snowy weather can cause a car accident. The accident is an emergent impact of the snow: the snow leads to slick roads and people driving focused more on retaining control. As an indirect result, two cars can end up in an accident. In addition to recognizing the snow as the origin of the accident, we could also say that the drivers steering their vehicles into each other was the cause, or that inability to see clearly was the cause, or any of many other proximate causes. That accident was not uniquely traceable to the snow as the cause.

3 The Fortune 500 lists the largest 500 companies in the world by revenues. The median number of employees for 10% of the Fortune 500 companies (every 10th) is 26,000. The mean of this sample is 200,000—although 2 of those 50 firms have over 2 million employees each. Meanwhile, "How Many Employees" (n.d.) estimates a mean average of 106,000 employees for all Fortune 500 companies.

4 Note: this measure is for farming alone, whereas if we referenced all food-related businesses, including food processing and restaurants, agriculture represents 5.5% of the U.S. economy (Economic Research Service 2017).

5 We do not judge Government on its reallocation of resources across consumers, leaving that important topic for other discussions.

6 Think tanks provide broad views of the future. However, such views are certainly biased. They are also as frequently driven by policy objectives as they are by the desire to provide neutral analysis. Increasing business consolidation also reinforces this trend.

7 Before the start of the Digital Technology Revolution in 1970, legislation either provided material detail about exactly how to execute the law, or enforcement was left to the states. The 1924 Oil Pollution Act, for example, explicitly prohibited the discharge of fuel oil into tidal waters ("Oil Pollution Act of 1924" 2017). Meanwhile, the 1955 Air Pollution Act provided national research and technical assistance while providing little in terms of national regulations.

8 The Clean Air Act also includes one specific, high-profile target designed to help achieve that goal: 90% reduction in auto emissions (EPA 2013).

9   An exemplary exception to bureaucratic flexibility in the Clean Air Act is the requirement that automobile exhaust must be reduced by over 90%. Frequently, legislators include similar exceptions in legislation (especially if lobbyists support them). Legislators certainly retain the ability to pass out pork.

10 The broad nature of such legislation also provides an extensive canvas within which special interests can lobby for specific clauses. Such special clauses can be included first in legislation, and later in regulations.

11 Businesses are expected to define, according to guidelines, the situations in which their vehicle will be safe. They then must provide proof that their vehicles meet those guidelines. This approach effectively flips the traditional approach in which standards, e.g., for safety belts, are identical for everyone. See Department of Transportation (2018), which is discussed in Shepardson (2017). The 2016 policy is also related to National Science and Technology Council (2016).

12 Efforts such as President Trump's orders to "eliminate two regulations for every new one," for example, are likely to be challenged in court: they may be inconsistent with what is explicitly written in the Clean Air Act (or other legislation).

13 Those viewing ethics of judges very highly fell from 53% to 45%. Congress from 15% to 8%, Senators from 19% to 12%, and Governors from 24% to 18%. Data on Congress and Senators is from 1976 to 2016; Judges and Governors are 1999 to 2016, due to data availability. It is noted that respect for state officeholders is flat, while respect for local officeholders has actually risen (37% to 58%.) ("Honesty/Ethics in Professions," n.d.).

14 Courtesy of Ruth Fisher, Quantaa. Based on "Code of Regulation Pages" (n.d.) and Crews (2014).

15 This estimate is based on research by Dawson and Seater (2013) and by Coffey, McLaughlin, and Peretto (2016). The cost is estimated as growth between 0.8% and 2% of the U.S. economy each year.

16 The original tunnel cost $48 million in 1920 currency ("Holland Tunnel" 2000), which is equivalent to $614 million today (Bureau of Labor Statistics, n.d.). The tunnel was built from 1920–1927 ("Holland Tunnel" 2018).

17 In a very different healthcare system in which citizens paid for the bulk of their own pharmaceuticals, such a regulation could be appropriate. However, that is not the state of current U.S. healthcare. See, for example, Sarnak et al. (2017).

18 Many countries put limits on the web pages available to residents. However, at best, such limitations are constrained in what they can do.

19 For further discussion see, for example, Crawford (2017).

20 The Institute of Electrical and Electronics Engineers (IEEE) is creating standards for Artificial Intelligence (IEEE, n.d. *Ethically Aligned Design*, 73–82).

21 Even where regulations exist, they are not necessarily enforced. The U.S. FCC, for example, just fined a toy maker a (trivial) amount for leaving data about millions of children publicly exposed. The company's activities broke laws, as well as the terms they promised customers. The Canadian equivalent of the FCC didn't bother with a fine at all (Coldewey 2018a).

22 See, for example, National Science and Technology Council (2016) and Kirchner (2017)

23 The total value of Bitcoin has grown to roughly of 1/3 a trillion dollars ("Market Capitalization," n.d.). Bitcoin has grown to the point where it may use as much electricity as powers Denmark, or at least as much as would come from one to three nuclear reactors (Mooney and Mufson 2017). Electricity is like air to Bitcoin; Bitcoin needs the electricity to operate. However, the electricity's use has no direct value for anyone else.

24 We leave analysis of public services, notably the military, to others.

25 Note: this data source for the 2015 number is different than those of earlier numbers. The consistency of the definitions is unknown.

26 Government so-called "leaning against the wind" to help prop up the U.S. economy during recessions explains little of the overall pattern. In other words, during recessions, many economists consider it reasonable for governments borrow money to prop up the economy, which they then pay off during booms.

27 The size of the deficit is often viewed in terms of U.S. GDP. That deficit is equivalent to 3.6% of U.S. GDP.

28 The only country with materially more debt than the U.S. is Japan. Japan's debt is over 300% of GDP, but that doesn't necessarily means the U.S. should, or even could incur that much debt. It's unlikely Japan will ever repay its debt (see, for example, Greenan and Weinstein [2017]); if the U.S. ever became completely unable to repay its debt, it would have unique challenges associated with its Congressional approval of increasing borrowing levels, maintaining the reserve currency for the world, having China own a material portion of U.S. debt, and being a net lender to the world. All of these, and other aspects, leave the prospect of never being able to repay U.S. debt troubling.

29 This debt is more like credit card debt than it is mortgage debt. With mortgage debt, we receive ongoing benefits from the home for which we are paying. Interest debt is paying for past consumption—which is what we are doing now.

CHAPTER 5

1 This is the cost difference at purchasing price parity.

2 For example, an "Artificial Intelligence in Business" meet-up, with presentations, beer, and pizza (Wallace 2017).

3 Based on an actual conversation, adjusted to hide the specifics of the company and product.

4 As discussed in Chapter 5, a restriction is a requirement to do or not to do something. E.g., anyone with a bank account more than $10,000 must list that account on Form 582 D.

5 Note: only low-sugar cereal is advertised because regulations limit the advertising of high-sugar cereals (Federal Trade Commission 2012).

6   Many of the most healthful foods such as fruits and vegetables may be more challenging to advertise as their brands are weaker. For example, Dole, Chiquita, and Ocean Spray have invested materially in their fruit-based brands. However, they are the exception. Many other reasonably healthful products from bread to ravioli to apple juice have strong brands and could be advertised. Yet, as the data above suggests, such products receive a small sliver of the advertising dollars spent today.

7   Patients may have been sent to the ER if they needed care that would take time—even if the critical care equipment of the ER wasn't needed; the example, a few hours of assistance to reduce dangerously high blood pressure.

8   This chart excludes Turkey, as Turkey's spending and life expectancy is materially below that of the other OECD countries.

9   A study of hospital readmission rates (Barnett, Hsu, and McWilliams 2015) found that the majority of differences in hospital readmission rates depend on patient socioeconomic factors—and only a minority depends on differences due to the hospital themselves. In other words, differences among patient Cultures and ability to pay for treatment have a greater impact on whether or not patients end up sick and readmitted to the hospital than do differences among specific hospital actions.

10 The following figure (5.16) illustrates the housing dynamics in greater detail: the shift from the first column (A*) to the second column (B*) is the immediate impact of a local economy boom. People want to move into a booming region, so the value of housing rises. As new housing is built, the house price decreases, to C*. The total cost of a home falls from B* to C*. Housing in such areas remains more expensive than housing in economically failing areas (A*). However, any area with a failing housing market—in which more housing is not built, retains prices that are even higher (B*).

Impact of Regional Success on Housing Prices

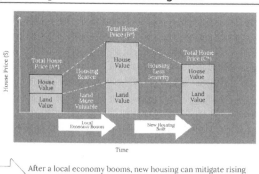

After a local economy booms, new housing can mitigate rising home prices.

Figure Endnote #10: A local economy boom should lead to housing price growth that cools off after needed housing is built.

11 Some may be concerned that we are addressing grade schools in the Markets section, when most grade schools are government operated. You could certainly compose a sound argument that grade schools would be better placed in Government. At the same time, grade schools are mixed private, public, and charter. In many areas these different school models compete with each other. Until it becomes overwhelmingly obvious which funding model is best for grade schools, we leave them in the Markets section.

12 A complete statement of purpose of education is beyond the scope of this document.

13 A grade of 'D' probably is fair: for reference, the U.S. now ranks 30th out of the 35 developed countries (OECD) in mathematics (National Center for Educational Statistics, n.d.).

14 Based on (Caplan 2018, 151). Approximately 10% of students can solve the challenge problem without the lesson problem. Only about 30% can solve the lesson problem with the challenge problem. If the students are explicitly told that the answers follow the same pattern, the percent raises to close to 75%.

15 Some students are getting a decent education, but they have too much debt to start new businesses after graduation (Ambrose, Cordell, and Ma 2015).

16 Recreation includes 12 hours socializing with friends, 11 hours using computers for fun, 6 hours watching television, 6 hours exercising, 5 hours on hobbies, and 3 hours on other forms of entertainment.

17 Net take-home pay adds all employment wages and then takes out taxes. It adds back in any subsidies, such as Government-subsidized rent or the Government paying the electric bill.

18 Loosely based on Katz (2015).

19 Chart created by Ruth Fisher of Quantaa: http://www.quantaa.com.

20 The upcoming data features "Total Factor Productivity," which takes into account changes in both labor and capital (physical investment). This example includes only productivity of labor, to illustrate the concept.

21 According to "GDP Per Capita" (2017), U.S. per capita GDP is $57K per person per year. 35% of that is almost $20K. If we include five more years of inflation, we would be over $20K.

22 Recently developed machine learning tools are spectacular. It is possible that productivity may, in fact, accelerate as such tools are implemented. However, the question is whether the gap between potential growth and actual growth will narrow. For the more technical reader, this next section's discussion is focused on that gap.

23 The Qin Emperor also was horrific in many ways (Eno 2010).

CHAPTER 6

1   Tools to understand the world as an interconnected, complex system go by many names: systems science (Ackoff 1999), complexity theory (Chu, Strand, and Fjelland 2003), and network analysis (Easley and Kleinberg 2010) are three common ones. Also noteworthy is decision-making under uncertainty. Each of these toolboxes share perspectives related to inter-connected decisions.

2   This statement is loosely based on differences in levels of sugar and antioxidants described by Watts (2013).

3   Semantics is the branch of logic concerned with meaning. Through semantics, each term and the overall structure of a statement can be precisely defined. For example, the statement, "Black cat ran" can be precisely defined through provision of definitions of black and cat, the operation ran, and the syntax of the language. Essentially, through such a defining process a world is defined. Through such precise definition, a computer or mathematician can uniquely confirm or reject that an equally-precisely defined scenario qualifies as "Black cat ran."

In principle, any logical statement can be expressed semantically precisely by defining terms and syntax. While most software is not semantically precisely defined (Mosses 2006), a vibrant academic literature exists on semantically precise computer programming. See, for example, Politz et al. (2013). Applications extend beyond computers, however. In principle, any logical statement can be semantically precisely communicated.

4   Such checking is done via mathematical proof. Once again, see Politz et al. (2013) for a related discussion.

5   This estimate is based on approximately $100,000 spent on fake news advertisements (Glum 2017) and a $64 billion total Russian government spending (Trading Economics, n.d.) converted to U.S. dollars based on the Google recorded exchange rate on April 24, 2018.

BIBLIOGRAPHY

Download the full bibliography at RodWallacePhD.com. The following is a sample:

AAPC (American Academy of Professional Coders). n.d. "ICD-10 Implementation." AAPC. Accessed January 29, 2018. https://www.aapc.com/icd-10/icd-10-implementation.aspx.

Aaron, Daniel G., and Michael B. Siegel. 2017. "Sponsorship of National Health Organizations by Two Major Soda Companies." *American Journal of Preventive Medicine* 52, no. 1: 20–30. https://doi.org/10.1016/j.amepre.2016.08.010.

ACCC (Australian Competition and Consumer Commission). n.d. "Scam Statistics." ACCC *ScamWatch*. Accessed February 20, 2018. https://www.scamwatch.gov.au/about-scamwatch/scam-statistics.

Ackoff, Russell L. 1999. *Re-Creating the Corporation: A Design of Organizations for the 21st Century*. New York: Oxford University Press.

Adams, Dominic. n.d. "Faces of Flint: Nicolas Carr." *MLive Media Group*. Accessed January 25, 2018. http://www.mlive.com/news/index.ssf/page/faces_of_flint_nicholas_carr.html.

Adams, Mike. 2015. "Isn't It Obvious? If Operation Jade Helm Were Happening in Any Other Country, It Would Be Immediately Labeled a Military Drill for Martial Law." *Natural News*. Posted March 30, 2015. https://www.naturalnews.com/049180_Operation_Jade_Helm_military_drill_martial_law.html.

AFSP (American Foundation for Suicide Prevention). n.d. "Facts About Suicide." AFSP. Accessed February 20, 2018. https://www.mobap.edu/wp-content/uploads/2013/01/afsp_facts_about_suicide_2007.pdf.

Agresti, James D. 2018. "National Debt Facts." *Just Facts*. Updated January 6, 2018. http://www.justfacts.com/nationaldebt.asp.

AI Now Institute. n.d. "AI Now: A Research Institute Examining the Social Implications of Artificial Intelligence." AI *Now Institute, New York University*. Accessed February 16, 2018. https://ainowinstitute.org.

Akpan, Nsikan. 2015. "The Race for the Unbreakable Password Is Almost Over." PBS *(Public Broadcasting Service)*. Posted August 27, 2015. http://www.pbs.org/newshour/updates/unbreakable-quantum-password/.

Alasaarela, Mikko. 2017. "The Rise of Emotionally Intelligent AI." *Machine Learnings*. Posted October 9, 2017. https://machinelearnings.co/the-rise-of-emotionally-intelligent-ai-fb9a814a630e.

Alexander, Scott. 2014. "The Toxoplasma of Rage." *Slate Star Codex*. Posted December 17, 2014. http://slatestarcodex.com/2014/12/17/the-toxoplasma-of-rage/.

Algedi, Markab. 2017. "'Smart City' in China Uses AI to Track Every Movement of Citizens, Enforce Laws." *Mind Unleashed*. Posted November 1, 2017. http://themindunleashed.com/2017/11/smart-city-china-uses-ai-track-every-movement-citizens-enforce-laws.html.

Amadeo, Kimberly. 2017a. "Current U.S. Federal Government Tax Revenue." *Balance*. Updated July 7, 2017. https://www.thebalance.com/current-u-s-federal-government-tax-revenue-3305762.

Amadeo, Kimberly. 2017b. "The Rising Cost of Health Care by Year and Its Causes." *Balance*. Updated October 26, 2017. https://www.thebalance.com/causes-of-rising-healthcare-costs-4064878.

Amadeo, Kimberly. 2018. "US GDP by Year Compared to Recessions and Events." *Balance*. Updated August 2, 2018. https://www.thebalance.com/us-gdp-by-year-3305543.

Ambrose, Brent W., Larry Cordell, and Shuwei Ma. 2015. " The Impact of Student Loan Debt on Small Business Formation." Working Paper No. 15-26, Research Department, Federal Reserve Bank of Philadelphia. https://www.philadelphiafed.org/-/media/research-and-data/publications/working-papers/2015/wp15-26.pdf.

American Board of Medical Specialties. n.d. "Specialty and Subspecialty Certificates." *American Board of Medical Specialties*. Accessed February 9, 2018. http://www.abms.org/member-boards/specialty-subspecialty-certificates/.

American Diabetes Association. n.d. "The Staggering Cost of Diabetes." Infographic, *American Diabetes Association*. Accessed January 24, 2018. http://main.diabetes.org/dorg/images/infographics/adv-cost-of-diabetes.gif.

American Society of Civil Engineers. 2017a "2017 Infrastructure Grades." Infographic. March 9, 2017. https://www.infrastructurereportcard.org/wp-content/uploads/2016/10/Grades-Chart.png.

American Society of Civil Engineers. 2017b "2017 Infrastructure Report Card: Aviation." March 9, 2017. https://www.infrastructurereportcard.org/wp-content/uploads/2017/01/Aviation-Final.pdf.

American Society of Civil Engineers. 2017c "2017 Infrastructure Report Card: Inland Waterways." March 9, 2017. https://www.infrastructurereportcard.org/wp-content/uploads/2017/01/Inland-Waterways-Final.pdf.

American Society of Civil Engineers. 2017d "2017 Infrastructure Report Card: Roads." March 9, 2017. https://www.infrastructurereportcard.org/wp-content/uploads/2017/01/Roads-Final.pdf.

Anderson, Monica, and Andrea Caumont. 2014. "How Social Media Is Reshaping News." *Pew Research Center*. Posted September 24, 2014. http://www.pewresearch.org/fact-tank/2014/09/24/how-social-media-is-reshaping-news/.

Andrews, Robin. 2016. "How Misinformation Spreads on the Internet." IFL *Science*. Posted January 6, 2016. http://www.iflscience.com/technology/facebook-echo-chambers-help-spread-and-reinforce-misinformation/.

Andris, Clio, David Lee, Marcus J. Hamilton, Mauro Martino, Christian E. Gunning, and John Armistead Selden. 2015. "The Rise of Partisanship and Super-Cooperators in the U.S. House of Representatives." *PLoS ONE* 10, no. 4: e0123507. https://doi.org/10.1371/journal.pone.0123507.

Animal and Plant Health Inspection Service. 2017a. "National Wildlife Disease Program (NWDP)." *United States Department of Agriculture.* Updated August 14, 2017. https://www.aphis.usda.gov/aphis/ourfocus/wildlifedamage/programs/nwrc/nwdp.

Animal and Plant Health Inspection Service. 2017b. "Vultures." *United States Department of Agriculture.* Updated August 16, 2017. https://www.aphis.usda.gov/aphis/ourfocus/wildlifedamage/operational-activities/sa_vultures/ct_vultures.

Animal Legal Defense Fund. n.d. "Animal Fighting Case Study: Michael Vick." *Animal Legal Defense Fund.* Accessed January 23, 2018. http://aldf.org/resources/laws-cases/animal-fighting-case-study-michael-vick/.

Anthony, Sebastian. 2014. "The Apollo 11 Moon Landing, 45 Years On: Looking Back at Mankind's Giant Leap." *Extreme Tech.* Posted July 21, 2014. https://www.extremetech.com/extreme/186600-apollo-11-moon-landing-45-years-looking-back-at-mankinds-giant-leap.

Anthony, Scott D., S. Patrick Viguerie, Evan I. Schwartz and John Van Landeghem. 2018. *2018 Corporate Longevity Forecast: Creative Destruction is Accelerating.* Executive Summary. Boston, MA: Innosight. https://www.innosight.com/insight/creative-destruction/.

Appel, Helmut, Alexander L. Gerlach, and Jan Crusius. 2016. "The Interplay Between Facebook Use, Social Comparison, Envy, and Depression." *Current Opinion in Psychology* 9:44–49. https://doi.org/10.1016/j.copsyc.2015.10.006.

"Apple II Series." 2018. *Wikipedia.* Updated February 10, 2018. https://en.wikipedia.org/wiki/Apple_II_series.

Arthur, W. Brian. 2009. *The Nature of Technology: What It Is and How It Evolves.* New York: Free Press.

"Artificial General Intelligence." 2018. *Wikipedia.* Updated March 23, 2018. https://en.wikipedia.org/wiki/Artificial_general_intelligence.

"Ashley Madison Data Breach." 2017. *Wikipedia.* Updated December 28, 2017. https://en.wikipedia.org/wiki/Ashley_Madison_data_breach.

"Asilomar AI Principles." n.d. *Beneficial AI Conference,* Asilomar, Pacific Grove, CA, January 5–8, 2017. Accessed January 30, 2018. https://futureoflife.org/ai-principles/.

Associated Press. 2017. "5-Year-Old Boy Seriously Injured After Pit Bull Drags Him Across Yard." *New York Post.* Posted September 23, 2017. http://nypost.com/2017/09/23/5-year-old-boy-seriously-injured-after-pit-bull-drags-him-across-yard/.

Athalye, Anish. 2017. "AI Image Recognition Fooled by Single Pixel Change." *BBC (British Broadcasting Corporation)* online. Posted November 3, 2017. http://www.bbc.com/news/technology-41845878.

Ault, Alicia. 2018. "Survey: 42% of Physician Report Burnout, Some Cite Depression." *Medscape.* Posted January 17, 2018. https://www.medscape.com/viewarticle/891411.

ENDNOTES

Aviva. 2017. *Aviva Home Report: Home Skills and Household Repairs*. London: Aviva.

Baetjer, Howard. 2016. "The Welfare Cliff and Why Many Low-Income Workers Will Never Overcome Poverty." *LearnLiberty*. Posted August 24, 2016. http://www.learnliberty.org/blog/the-welfare-cliff-and-why-many-low-income-workers-will-never-overcome-poverty/.

Bailey, Scott. 1956. "They Build Model T's" *Boys' Life*. (March 1956):16, 71.

Barclay, Eliza. 2016. "Scientists Are Building a Case for How Food Ads Make Us Overeat." *National Public Radio* online. Updated February 1, 2016. http://www.npr.org/sections/thesalt/2016/01/29/462838153/food-ads-make-us-eat-more-and-should-be-regulated.

Barnes, Taylor. 2009. "America's 'Shadow Economy' Is Bigger Than You Think — and Growing." *Christian Science Monitor*. Posted November 12, 2009. https://www.csmonitor.com/Business/2009/1112/americas-shadow-economy-is-bigger-than-you-think-and-growing.

Barnett, Michael L., John Hsu, and J. Michael McWilliams. 2015. "Patient Characteristics and Differences in Hospital Readmission Rates." *JAMA Internal Medicine* 175, no. 11: 1803–12. http://dx.doi.or/10.1001/jamainternmed.2015.4660.

Barstow, David, and Robin Stein. 2005. "Under Bush, a New Age of Prepackaged TV News." *New York Times* online. Posted March 13, 2005. http://www.nytimes.com/2005/03/13/politics/under-bush-a-new-age-of-prepackaged-tv-news.html.

Basset, Angela. 2017. Posting on *Twitter* by Laura E. Davis. June 13, 2017. https://twitter.com/lauraelizdavis/status/874685885408833536.

Bedard, Paul. 2016. "Surprise: Washington Post Profited Off Trump Election." *Washington Examiner*. Posted December 16, 2016. http://www.washingtonexaminer.com/surprise-washington-post-profited-off-trump-election/article/2609820.

Belkin, Douglas. 2017. "Exclusive Test Data: Many Colleges Fail to Improve Critical-Thinking Skills." *Wall Street Journal* online. Posted June 5, 2017. https://www.wsj.com/articles/exclusive-test-data-many-colleges-fail-to-improve-critical-thinking-skills-1496686662.

Bellini, Jason. 2018. "Why 'Deaths of Despair' May Be a Warning Sign for America — Moving Upstream." *Wall Street Journal* online. Posted February 27, 2018. https://www.wsj.com/articles/why-deaths-of-despair-may-be-a-warning-sign-for-america-moving-upstream-1519743601.

Bellis, Mary. 2018. "The Invention of Radio Technology." *ThoughtCo*. Updated March 2, 2018. https://www.thoughtco.com/invention-of-radio-1992382.

Belluz, Julia. 2015. "How Mass Breast Cancer Screening Failed to Diagnose Serious Cases, in One Chart." *Vox*. Updated October 29, 2015. https://www.vox.com/2015/10/28/9631500/does-mammography-work.

Benet, William E. n.d. "Psychometrics." *Assessment Psychology Online*. Accessed January 23, 2018. http://www.assessmentpsychology.com/psychometrics.htm.

Berard, Yamil. 2014. "Fort Worth District Audit Finds $2.7 Million in Unneeded Technology." *Star-Telegram* online. Updated July 134, 2014. http://www.star-telegram.com/news/local/community/fort-worth/article3865141.html.

Berman, Jillian. 2016. "America's Growing Student-Loan-Debt Crisis." *Market Watch*. Posted January 19, 2016. https://www.marketwatch.com/story/americas-growing-student-loan-debt-crisis-2016-01-15.

Bolukbasi, Tolga, Kai-Wei Chang, James Zou, Venkatesh Saligrama, and Adam Kalai. 2016. "Man Is to Computer Programmer as Woman Is to Homemaker? Debiasing Word Embeddings." *Advances in Neural Information Processing Systems* 29, ed. D. D. Lee, M. Sugiyama, U. V. Luxburg, I. Guyon, and R. Garnett, 4349–57. Red Hook, NY: Curran Associates. https://papers.nips.cc/paper/6228-man-is-to-computer-programmer-as-woman-is-to-homemaker-debiasing-word-embeddings.

Bond, Robert M., Christopher J. Fariss, Jason J. Jones, Adam D. I. Kramer, Cameron Marlow, Jaime E. Settle, and James H. Fowler. 2012. "A 61-Million-Person Experiment in Social Influence and Political Mobilization." *Nature* 489:295–98. https://doi.org/10.1038/nature11421.

Bonnington, Christina. 2014. "A Smartwatch that Nudges You to Meet Your Fitness Goals." *Wired* online. Posted September 30, 2014. https://www.wired.com/?p=1578513.

Booth, Robert. 2014. "Facebook Reveals News Feed Experiment to Control Emotions." *Guardian* online. Posted June 30, 2014. https://www.theguardian.com/technology/2014/jun/29/facebook-users-emotions-news-feeds.

Borgatti, Stephen P. n.d. "Differences in Cultures." Lecture notes for MB022, "Organizational Behavior," 2001, Boston College. Accessed February 15, 2018. http://www.analytictech.com/mb021/cultural.htm.

Boscamp, Emi. 2017. "Dining Etiquette from Around the World." *Huffington Post* online. Updated December 6, 2017. http://www.huffingtonpost.ca/entry/dining-etiquette-around-the-world_n_3567015.

Bowling, Lauren. 2017. "How to Move to a New City with No Money and No Prospects." *Financial Best Life*. Updated July 5, 2017. https://financialbestlife.com/how-to-move-with-no-money/.

Bowman, Sue. 2013. "Impact of Electronic Health Record Systems on Information Integrity: Quality and Safety Implications." *Perspectives in Health Information Management* 10:1c. https://www.ncbi.nlm.nih.gov/pmc/articles/PMC3797550/.

Braga, Matthew. 2018. "How One Researcher Harvested Data from 50 Million People — and Facebook Was Designed to Help." CBC (*Canadian Broadcasting Corporation*) online. Updated April 19, 2018. https://www.cbc.ca/news/technology/facebook-cambridge-analytica-friends-api-by-design-1.4583337.

Braithwaite, Jeffrey, Robert L. Wears, and Erik Hollnagel. 2016. *Resilient Health Care, Volume 3: Reconciling Work-as-Imagined and Work-as-Done*. New York: CRC Press.

Brandt, Michelle. 2012. "Little Evidence of Health Benefits from Organic Foods, Study Finds." *Stanford Medicine*. Posted September 3, 2012. https://med.stanford.edu/news/all-news/2012/09/little-evidence-of-health-benefits-from-organic-foods-study-finds.html.

Bratskeir, Kate. 2015. "Restaurant Meals Are Just as Unhealthy as Fast Food." *Huffington Post*. Updated July 24, 2015. https://www.huffingtonpost.com/entry/fancy-restaurant-meals-may-be-just-as-unhealthy-as-big-macs-and-fries_ us_55ad10e3e4b065dfe89ec82c.

Bratton, Benjamin H. 2016. *The Stack: On Software and Sovereignty.* Cambridge, MA: MIT Press. https://doi.org/10.7551/mitpress/9780262029575.001.0001.

Brenan, Megan. 2017. "Nurses Keep Healthy Lead as Most Honest, Ethical Profession." *Gallup.* Posted December 26, 2017. http://news.gallup.com/poll/224639/nurses-keep-healthy-lead-honest-ethical-profession.aspx.

Brindle, James. 2017. "Something Is Wrong on the Internet." *Medium.* Posted November 6, 2017. https://medium.com/@jamesbridle/something-is-wrong-on-the-internet-c39c471271d2.

Brint, Steven, and Allison M. Cantwell. 2010. "Undergraduate Time Use and Academic Outcomes: Results from the University of California Undergraduate Experience Survey 2006." *Teachers College Record* 112, no. 9: 2441–70. http://www.higher-ed2000.ucr.edu/Publications/Brint%20and%20Cantwell%202008.pdf.

Britz, Denny. 2017. "Engineering Is the Bottleneck in (Deep Learning) Research." *Denny's Blog.* Posted January 17, 2017. http://blog.dennybritz.com/2017/01/17/engineering-is-the-bottleneck-in-deep-learning-research/.

Brunnermeier, Markus K., and Martin Oehmke. 2012. "Bubbles, Financial Crises, and Systemic Risk." Cambridge, MA: National Bureau of Economic Research (NBER). NBER Working Paper Series #18398. https://doi.org/10.3386/w18398.

Bryant, Martin. 2011. "20 Years Ago Today, the World Wide Web Opened to the Public." *NextWeb.* Posted August 6, 2011. https://thenextweb.com/insider/2011/08/06/20-years-ago-today-the-world-wide-web-opened-to-the-public/.

Brynjolfsson, Erik, Daniel Rock, and Chad Syverson. 2017. "Artificial Intelligence and the Modern Productivity Paradox: A Clash of Expectations and Statistics," Cambridge, MA: National Bureau of Economic Research (NBER). NBER Working Paper Series #24001. https://doi.org/10.3386/w24001.

Bryon, Ellen. 2017. "America's Retailers Have a New Target Customer: The 26-Year-Old Millennial." *Wall Street Journal.* Posted October 9, 2017. https://www.wsj.com/articles/americas-retailers-have-a-new-target-customer-the-26-year-old-millennial-1507559181.

Bull, Matthew J., and Nigel T. Plummer. 2014. "Part 1: The Human Gut Microbiome in Health and Disease." *Integrative Medicine: A Clinician's Journal* 13, no. 6: 17–22. https://www.ncbi.nlm.nih.gov/pmc/articles/PMC4566439/.

Bunim, Juliana. 2013. "Quantity of Sugar in Food Supply Linked to Diabetes Rates." *University of California, San Francisco.* Posted February 27, 2013. https://www.ucsf.edu/news/2013/02/13591/quantity-sugar-food-supply-linked-diabetes-rates.

Bureau of Labor Statistics. n.d. "CPI Inflation Calculator." *Bureau of Labor Statistics, United States Department of Labor.* Accessed January 25, 2018. https://www.bls.gov/data/inflation_calculator.htm.

Bush, George W. 2005. News conference at the White House, Washington, DC, Wednesday, March 16, 2005. Transcript by FDCH E-Media, *Washington Post.* http://www.washingtonpost.com/wp-dyn/articles/A40191-2005Mar16_5.html. Video coverage available on YouTube, https://www.youtube.com/watch?v=4OJab6IUaU8.

California Department of Education. 2000. *English–Language Arts Content Standards for California Public Schools – Kindergarten Through Grade Twelve.* Sacramento, CA: California Department of Education. https://www.cde.ca.gov/be/st/ss/documents/elacontentstnds.pdf.

California Tax Data. n.d. "What Is Proposition 13?" *California Tax Data, National Tax Data Inc.* Accessed January 29, 2018. https://www.californiataxdata.com/pdf/Prop13.pdf.

Campbell-Dollaghan, Kelsey. 2015. "The Forgotten Story of NYC's First Power Grid." *Gizmodo.* Posted January 26, 2015. https://gizmodo.com/the-forgotten-story-of-nycs-first-power-grid-1681857054.

Caplan, Bryan. 2018. *The Case Against Education: Why the Education System Is a Waste of Time and Money.* Princeton, NJ: Princeton University Press.

Cargill. 2017. "More: The 2016/2017 Cargill Cocoa Promise Global Summary Report." *Cargill.* https://www.cargill.com/doc/1432099950824/cargill-cocoa-promise-report-2016-17.pdf.

Carp, Jeremy. 2015. "Should Congress Simplify Regulation of Higher Education?" *Regulatory Review.* Posted September 29, 2015. https://www.theregreview.org/2015/09/29/carp-simplifying-higher-ed-regulation/.

Carroll, John. 2017. "FDA Experts Offer a Unanimous Endorsement for Pioneering Gene Therapy for Blindness." *Science.* Posted October 13, 2017. http://www.sciencemag.org/news/2017/10/fda-experts-offer-unanimous-endorsement-pioneering-gene-therapy-blindness.

Cascone, Susan. 2017. "Artificial Intelligence Can Now Spot Art Forgeries by Comparing Brush Strokes." *Artnet.* Posted November 21, 2017. https://news.artnet.com/art-world/artificial-intelligence-art-authentication-1156037.

Cassidy, John. 2015. "The Day Trader and the Flash Crash: Unanswered Questions." *New Yorker* online. Posted April 23, 2015. https://www.newyorker.com/news/john-cassidy/the-day-trader-and-the-flash-crash-unanswered-questions.

CDC (Centers for Disease Control and Prevention). 2016. "1 in 3 Adults Don't Get Enough Sleep." *CDC Newsroom.* Updated February 16, 2016. https://www.cdc.gov/media/releases/2016/p0215-enough-sleep.html.

CDC (Centers for Disease Control and Prevention). 2017. "Chronic Disease Overview." *CDC.* Updated June 28, 2017. https://www.cdc.gov/chronicdisease/overview/index.htm.

Chamorro-Premuzic, Tomas. 2014. "How the Web Distorts Reality and Impairs Our Judgement Skills." *Guardian* online. Posted May 13, 2014. https://www.theguardian.com/media-network/media-network-blog/2014/may/13/internet-confirmation-bias.

Chapman University. 2017. "America's Top Fears 2017: Chapman University Survey of American Fears." *Chapman University.* Posted October 11, 2017. https://blogs.chapman.edu/wilkinson/2017/10/11/americas-top-fears-2017/.

Chappell, Bill. 2015. "Affair-Enabling Website Ashley Madison Is Compromised by Hackers." *National Public Radio* online. Posted July 20, 2015. https://www.npr.

org/sections/thetwo-way/2015/07/20/424637005/affair-enabling-website-ashleymadison-is-compromised-by-hackers.

Chayko, Mary. 2016. *Superconnected: The Internet, Digital Media, and Techno-Social Life*. Thousand Oaks, CA: Sage.

"Choosing the Right Estimator." n.d. *Scikit-learn developers*. Accessed February 12, 2018. http://scikit-learn.org/stable/tutorial/machine_learning_map/.

Chow, Yan, and Molly Coye. 2013. "Rethinking Healthcare Delivery: Choosing New Technologies." 2013 Healthcare Innovation Summit, moderated by Arnold Milstein and Stefanos Zenios. *YouTube*. Posted April 29, 2013. https://youtu.be/f8mCxf0KGrE.

Chrisomalis, Stephen. n.d. "Science and Studies." *Phrontistery*. Accessed February 9, 2018. http://phrontistery.info/sciences.html.

Christakis, Dimitri A. 2010. "Internet Addiction: A 21st Century Epidemic?" BMC *Medicine* 8, no. 61. http://doi.org/10.1186/1741-7015-8-61.

Chu, Dominique, Roger Strand, and Ragnar Fjelland. 2003. "Theories of Complexity." *Complexity* 8, No. 3: 19–30. http://dx.doi.org/10.1002/cplx.10059.

Cifu, Adam. 2016. "Adam Cifu on Ending Medical Reversal." Interview by Russell Roberts, February 15, 2016. *EconTalk* podcast, 1:04:49. http://www.econtalk.org/archives/2016/02/adam_cifu_on_en.html.

Cilluffo, Anthony, Abigail Geiger, and Richard Fry. 2017. "More U.S. Households Are Renting than at Any Point in 50 Years." *Pew Research Center*. Posted July 19, 2017. http://www.pewresearch.org/fact-tank/2017/07/19/more-u-s-households-are-renting-than-at-any-point-in-50-years/.

Clark, Leslie. 2006. "In Memoriam 1916–2009." Eulogy for Walter Cronkite. PBS *(Public Broadcasting Service)*. Posted July 26, 2006. http://www.pbs.org/wnet/americanmasters/walter-cronkite-about-walter-cronkite/561/.

CMS (Centers for Medicare and Medicaid Services, United States). 2018. "National Health Expenditure Data: Historical." CMS, Baltimore, MD. Updated January 8, 2018. https://www.cms.gov/research-statistics-data-and-systems/statistics-trends-and-reports/nationalhealthexpenddata/nationalhealthaccountshistorical.html.

"The Code of Hammurabi". Circa 1780 B.C.E. Translated by L. W. King, *Encyclopaedia Britannica*, eleventh edition. *Internet Sacred Text Archive*. Accessed May 9, 2018. http://www.sacred-texts.com/ane/ham/index.htm.

"Code of Regulation Pages, 1970–2003." n.d. Graph. *Heritage Foundation*. Accessed January 23, 2018. https://www.heritage.org/sites/default/files/%7E/media/images/reports/bg1801chart2.jpg.

Coffey, Bentley, Patrick A. McLaughlin, and Pietro Peretto. 2016. "The Cumulative Cost of Regulations." Working paper, *Mercatus Center, George Mason University*. https://www.mercatus.org/publication/cumulative-cost-regulations.

"Cognitive Dissonance." 2018. *Wikipedia*. Updated February 14, 2018. https://en.wikipedia.org/wiki/Cognitive_dissonance.

Cohen, Paula. 2015. "How the 'Chocolate Diet' Hoax Fooled Millions." CBS *News*. Posted May 29, 2015. http://www.cbsnews.com/news/how-the-chocolate-diet-hoax-fooled-millions/.

Coldewey, Devin. 2018a. "After Breach Exposing Millions of Parents and Kids, Toymaker VTech Handed a $650K Fine by FTC." *Yahoo! Finance.* Posted January 8, 2018. https://finance.yahoo.com/news/breach-exposing-millions-parents-kids-184726572.html.

Coldewey, Devin. 2018b. "Google Temporarily Bans Addiction Center Ads Globally Following Exposure of Seedy Referral Deals." *TechCrunch.* Posted January 12, 2018. https://techcrunch.com/2018/01/12/google-temporarily-bans-addiction-center-ads-globally-following-exposure-of-seedy-referral-deals/.

Commodity Futures Trading Commission. 2017. "17 CFR Part 1: RIN 3038-AE62: Retail Commodity Transactions Involving Virtual Currency." *Federal Resister* 82, no. 243: 60335–341. Washington, DC: Office of the Federal Register. http://www.cftc.gov/idc/groups/public/@lrfederalregister/documents/file/2017-27421a.pdf.

"Communications Decency Act." 2017. *Wikipedia.* Updated December 7, 2017. https://en.wikipedia.org/wiki/Communications_Decency_Act.

"Company Facts." n.d. *Walmart.* Accessed January 28, 2018. https://corporate.walmart.com/newsroom/company-facts.

Computer History Museum. 2015. "1970: Semiconductors Compete with Magnetic Cores". *Computer History Museum.* Updated August 24, 2015. http://www.computerhistory.org/storageengine/semiconductors-compete-with-magnetic-cores/.

Condliffe, Jamie. 2017. "DeepMind's New Ethics Team Wants to Solve AI's Problems Before They Happen." MIT *Technology Review.* Posted October 4, 2017. https://www.technologyreview.com/the-download/609044/deepminds-new-ethics-team-wants-to-solve-ais-problems-before-they-happen/.

Constine, Josh. 2018a. "Facebook Feed Change Sacrifices Time Spent and News Outlets for 'Well-Being'." *TechCrunch.* Posted January 11, 2018. https://techcrunch.com/2018/01/11/facebook-time-well-spent/.

Constine, Josh. 2018b. "Facebook's U.S. User Count Declines as it Prioritizes Well-Being." *TechCrunch.* Posted January 31, 2018. https://techcrunch.com/2018/01/31/facebook-time-spent/.

Consumer Reports. 2014. "What to Do When There Are Too Many Product Choices on the Store Shelves?" *Consumer Reports* online. Posted January 2014.

http://www.consumerreports.org/cro/magazine/2014/03/too-many-product-choices-in-supermarkets/index.htm.

Consumer Reports. 2015. "The Cost of Organic Food: A New Consumer Reports Study Reveals How Much More You'll Pay. Hint: Don't Assume That Organic Is Always Pricier." *Consumer Reports* online. Posted March 19, 2015. https://www.consumerreports.org/cro/news/2015/03/cost-of-organic-food/index.htm.

Copp, Tara. 2018. "The Death Toll for Rising Aviation Accidents: 133 Troops Killed in Five Years." *Military Times.* Posted April 8, 2018. https://www.militarytimes.com/news/your-military/2018/04/08/the-death-toll-for-rising-aviation-accidents-133-troops-killed-in-five-years/.

Council of Economic Advisers. 2016. *The Long-Term Decline in Prime-Age Male Labor Force Participation.* Washington, DC: Executive Office of the President of the United States of America. https://obamawhitehouse.archives.gov/sites/default/files/page/files/20160620_cea_primeage_male_lfp.pdf

Crawford, Chris. 2014. "One in Three Patients not Filling Prescriptions, Study Finds." *AAFP News* (American Academy of Family Physicians). Posted April 28, 2014. https://www.aafp.org/news/health-of-the-public/20140428nonadherencestudy.html.

Crawford, Kate. 2017. "The Trouble with Bias — NIPS 2017 Keynote — Kate Crawford #NIPS2017." Presentation at Neural Information Processing Systems Conference (NIPS), December 5, 2017. *The Artificial Intelligence Channel.* Posted December 10, 2017. https://www.youtube.com/watch?v=fMym_BKWQzk&app=desktop.

Crawford, Kate, and Ryan Calo. 2016. "There Is a Blind Spot in AI Research." *Nature* 538, no. 7625: 311–13. Posted online October 13, 2016. https://www.nature.com/news/there-is-a-blind-spot-in-ai-research-1.20805.

Creswell, Julie, and Sapna Maheshwari. 2017. "United Grapples with PR Crisis Over Videos of Man Being Dragged Off Plane." *New York Times* online. Posted April 11, 2017. https://www.nytimes.com/2017/04/11/business/united-airline-passenger-overbooked-flights.html.

Crews, Clyde Wayne. 2014. "New Data: Code of Federal Regulations Expanding, Faster Pace under Obama." *Competitive Enterprise Institute.* Posted March 17, 2014. https://cei.org/blog/new-data-code-federal-regulations-expanding-faster-pace-under-obama.

Crooks, Glenna. 2018. *The NetworkSage: Realize Your Network Superpower.* Bloomington, IN: iUniverse.

Crowe, Jonathan. 2016. "Phishing by the Numbers: Must-Know Phishing Statistics 2016." *Barkly.* Posted July 2016. https://blog.barkly.com/phishing-statistics-2016.

Crowell, Maddy. 2015. "How Computers Are Getting Better at Detecting Liars." *Christian Science Monitor* online. Posted December 12, 2015. https://www.csmonitor.com/Science/Science-Notebook/2015/1212/How-computers-are-getting-better-at-detecting-liars.

Curtin, Sally C., Margaret Warner, and Holly Hedegaard. 2016. "Increase in Suicide in the United States, 1999–2014." *Centers for Disease Control and Prevention* (CDC), NCHS Data Brief No. 241, April 2016. https://www.cdc.gov/nchs/products/databriefs/db241.htm.

"Daddona, Nicole. 2016. *Buff Bernie: A Coloring Book for Berniacs.* Createspace Independent Publishing Platform.

Danny. 2018. "37 Mind Blowing YouTube Facts, Figures and Statistics — 2018." *MerchDope.* Posted April 26, 2018. https://merchdope.com/youtube-statistics/.

Davies, Dave. 2016. "Aging and Unstable: The Nation's Electrical Grid Is 'The Weakest Link.'" Interview with Gretchen Bakke, *Fresh Air,* National Public Radio. Transcript, August 22, 2016. https://www.npr.org/2016/08/22/490932307/aging-and-unstable-the-nations-electrical-grid-is-the-weakest-link.

Dawson, John W., and John J. Seater. 2013. "Federal Regulation and Aggregate

Economic Growth." *Journal of Economic Growth* 18, no. 2: 137–77. https://doi.org/10.1007/s10887-013-9088-y.

Delaney, John K. 2018. "It's Time for Washington to Start Working on Artificial Intelligence." *Telecrunch.* Posted January 17, 2018. https://techcrunch.com/2018/01/17/its-time-for-washington-to-start-working-onartificial-intelligence/.

Department of Transportation (United States). 2018. "USDOT Automated Vehicles Activities." 2018. *United States Department of Transportation.* Updated January 11, 2018. https://www.transportation.gov/AV.

Diamandis, Peter. 215. "Ray Kurzweil's Mind-Boggling Predictions for the Next 25 Years." *Singularity Hub.* Posted January 26, 2018. https://singularityhub.com/2015/01/26/ray-kurzweils-mind-boggling-predictions-for-the-next-25-years/.

Dignan, Larry. 2017. "The Great Data Science Hope: Machine Learning Can Cure Your Terrible Data Hygiene." ZDNet. Posted November 12, 2017. http://www.zdnet.com/article/the-great-data-science-hope-machine-learning-can-cure-your-terrible-data-hygiene/.

Dillian, Jared. 2017. "The Everything Bubble." *Mauldin Economics.* Posted June 22, 2017. http://www.mauldineconomics.com/the-10th-man/the-everything-bubble.

Dionne, E.J., Jr. 2014. "Where Goes the Neighborhood?" *Washington Post* online. Posted August 10, 2014. https://www.washingtonpost.com/opinions/ej-dionne-where-goes-the-neighbor...0/8a137cde-1f39-11e4-ae54-0cfe1f974f8a_story.html?utm_term=.aed72c8b832a.

Doherty, Tucker, and Jack Shafer. 2017. "The Media Bubble Is Worse than You Think." *Politco Magazine,* May/June 2017. http://www.politico.com/magazine/story/2017/04/25/media-bubble-real-journalism-jobs-east-coast-215048.

Douris, Constance. 2017. "The Bottom Line On Electric Cars: They're Cheaper to Own." *Forbes* online. Posted October 24, 2017. https://www.forbes.com/sites/constancedouris/2017/10/24/the-bottom-line-on-electric-cars-theyre-cheaper-to-own/#56385fa210b6.

Dowall, David E. 1982. "The Suburban Squeeze: Land-Use Policies in the San Francisco Bay Area." *Cato Journal* 2, no. 3: 709–33. https://object.cato.org/sites/cato.org/files/serials/files/cato-journal/1983/1/cj2n3-4.pdf.

Dreyfuss, Emily. 2017. "Want to Make a Lie Seem True? Say It Again. And Again. And Again." *Wired.* Posted February 11, 2017. https://www.wired.com/2017/02/dont-believe-lies-just-people-repeat/.

Drucker, Peter F. 1970. *Technology, Management, and Society.* New York: Routledge.

Drucker, Peter F. 1993. *The Ecological Vision: Reflections on the American Condition.* New Brunswick, NJ: Transaction Publishers.

Drutman, Lee, and Steven Teles. 2015. "Why Congress Relies on Lobbyists Instead of Thinking for Itself." *Atlantic* online. Posted March 10, 2015. https://www.theatlantic.com/politics/archive/2015/03/when-congress-cant-think-for-itself-it-turns-to-lobbyists/387295/.

"Duke Lacrosse Case." 2018. *Wikipedia*. Updated February 16, 2018. https://en.wikipedia.org/wiki/Duke_lacrosse_case.

Dunbar, Robin. 2011. *How Many Friends Does One Person Need? Dunbar's Number and Other Evolutionary Quirks*. London, UK: Faber and Faber.

Dunkelberg, William. 2016. "The Hidden Cost of Regulations." *Forbes*.

Posted July 12, 2016. https://www.forbes.com/sites/williamdunkelberg/2016/07/12/the-cost-of-regulations/#39a60a66c812.

Dunkelman, Marc J. 2014. *The Vanishing Neighbor: The Transformation of American Community*. New York: W.W. Norton & Company. Kindle Edition.

Duportail, Judith. 2017. "I Asked Tinder for My Data. It Sent Me 800 Pages of My Deepest, Darkest Secrets." *Guardian* online. Posted September 26, 2017. https://www.theguardian.com/technology/2017/sep/26/tinder-personal-data-dating-app-messages-hacked-sold.

Durden, Tyler. 2017a. "Home Depot Panics Over Millennials; Forced to Host Tutorials on Using Tape Measures, Hammering Nails." *ZeroHedge*. Posted October 10, 2017. http://www.zerohedge.com/news/2017-10-10/home-depot-panicked-over-millennials-forced-host-tutorials-using-tape-measures-hamme.

Durden, Tyler. 2017b. "Nasdaq Triggers Market-Wide Circuit-Breaker as AMZN 'Crashes' 87% After-Hours." *ZeroHedge*. Posted July 4, 2017. http://www.zerohedge.com/news/2017-07-03/nasdaq-triggers-market-wide-circuit-breaker-amzn-crashes-87-after-hours.

Durden, Tyler. 2017c. "It Is Seven Times More Difficult to Get a Flight Attendant Job at Delta than Enter Harvard." *ZeroHedge*. Posted October 24, 2017. http://www.zerohedge.com/news/2017-10-23/it-seven-time-more-difficult-get-flight-attendant-job-delta-enter-harvard.

Dvorak, Nate, and Bryant Ott. 2015. "A Company's Purpose Has to Be a Lot More than Words." *Gallup*. Posted July 28, 2015. http://news.gallup.com/businessjournal/184376/company-purpose-lot-words.aspx.

DOWNLOAD THE FULL
BIBLIOGRAPHY AT
RODWALLACEPHD.COM

INDEX

artificial intelligence 49, 50, 55, 87–88, 90, 108, 140, 195, 237–242, 310, 335

collaboration

as human endeavor 31–33, 35–37, 43–44, 68–69, 103, 151, 155, 158, 161–
162, 242–245, 272–274, 322–326, 352–353, 360, 369
in aligning use of D-Tech 6–7, 42, 85–86, 89–90, 90–
91, 99, 200, 259, 261, 368. See also technology, digital: outdated/legacy
systems

complexity

context, importance of understanding. See information: context
interconnectedness, 20, 24, 80–84, 177–178, 194, 199–201, 223–229, 238–
247, 253–254, 259–260
managing 25–36, 43–44, 80–82, 88–89, 165, 177–178, 191–193, 199–201, 204–
206, 225–229, 238–243, 257–259, 323, 360, 363–365, 366–371
rapid change 3, 16–21, 56–57, 88–89, 179–180, 196–197, 277

culture

cultural norms 99–100, 103, 111–112, 113, 149, 151, 152–153, 158–159, 263
social discourse 32, 99–100, 100–101, 103, 104, 109, 111, 112–
156, 158, 163, 265. See also society: storytelling

decision-making

influence of experts 42, 91–92, 174–175, 180–184, 185, 204–206
managing complexity. See complexity: managing

economy

concentration (monopoly) 183, 211–213, 316–317
inequality 35–36, 109–111, 159, 263–264, 287, 297–300
markets. See markets
productivity 35–38, 269, 295–298, 306–312

education

declining results and rising costs 214, 278–299
increasing specialization. See labor: specialization
use of digital technology in 3, 288, 291, 292–294

27342167R00227